THE LAST RUN

THE
LAST RUN

A TRUE STORY OF
RESCUE AND REDEMPTION
ON THE ALASKA SEAS

TODD LEWAN

CPL

HarperCollins*Publishers*

HarperCollins books may be purchased for educational, business, or sales promotional use. For information, please write: Special Markets Department, HarperCollins Publishers Inc., 10 East 53rd Street, New York, NY 10022.

FIRST EDITION

Designed by Elliott Beard

Map by John T. Sebastian

Printed on acid-free paper

Library of Congress Cataloging-in-Publication Data is available upon request.

ISBN 0-06-019648-3

04 05 06 07 08 ❖/RRD 10 9 8 7 6 5 4 3

To my mother,
who never stopped giving

ACKNOWLEDGMENTS

I have always liked searching for pieces to a story. There is something revitalizing about discovering a piece of information that no one else has, or, what is more likely, finding a piece that others have discovered but not considered in quite the same way. I suppose this is why I spent four years researching and writing this book. The rush of discovery.

Then again, it might just have been my fascination with the men who appear in this book. They, of course, are the fishermen and Coast Guardsmen who survived the events of January 30, 1998, and who endured one-on-one interviews, phone calls and e-mails many months after that night. It is bad enough going through so horrific a tempest; having someone stir it up afresh through repeated questioning and prodding cannot be pleasant.

I didn't expect any of them to remember so many details about that January night, and I certainly could not expect them to be so forthright about their pasts, their dreams, their demons, and yet they were. They shared some pretty personal stuff. To them, I can only offer my most sincere thanks.

Of course, there were others who helped me to collect and arrange the pieces of this story-puzzle, people who do not appear in the story but whose contributions were no less significant. They are numerous—so numerous that, sadly, I cannot list all of their names. Some, however, deserve a special mention.

To start, there was my editor at The Associated Press, Bruce DeSilva, who agreed to send me to Alaska for six weeks in September 1998 even though I hadn't the faintest idea what I was going to write about. Upon my return to New York, I told him and his assistant, Chris Sullivan, about the *La Conte* story, and I suggested writing it in the old-style, serialized format. The news cooperative, as far as anyone knew, had never run a serial on the news wire in the AP's 150-year history. But these two men pushed to get the story on the wire in five installments, no small feat at an agency that prides itself on brevity.

I am also deeply indebted to Larry Mussara, a retired Coast Guard helicopter pilot who first told me about the *La Conte* case while we fished on his boat over Labor Day weekend in 1998. It was Larry who talked me into writing a newspaper story about the rescue, and it was he who got me started by introducing me to officers at the District 17 headquarters in Juneau.

Others who offered unstinting help were Kent Lind, a biologist with the National Marine Fisheries Service in Juneau, who kindly allowed me to occupy his abode on more than one occasion; Ray Massey, the Coast Guard's spokesman at District 17 headquarters; George Bancroft, who explained the mechanics of the storm at the Marine Prediction Center in Camp Springs, Maryland; Ajay Mehta and Lieutenant Commander Paul Steward, who made clear the intricacies of emergency satellite beacons and the mainframe computer at Mission Control Center in Suitland, Maryland; Jim Jensen, director of the Yakutat police, who, along with Eli Hanlon and the Yakutat hospital staff, was invaluable in helping to reconstruct scenes at the end of the book; Dr. Michael T. Propst, John Bond, Alvin Ancheta, Walter MacFarlane, Dale Bivins, David Johnson and, in particular, David Hanson, all of the Alaska state crime lab in Anchorage, who made the opening chapters of the book possible. To all of you, my heartiest thanks.

In Kodiak, I had the pleasure of working with Sergeant Darlene Turner, chief of the station there, and two of her officers, Steve Hall and Tom Dunn. Without them, I would not have been able to re-create the book's opening scene. Dunn flew with me to Shuyak Island in a tiny, chartered plane through a blinding winter downpour and kept a keen (nervous, you could say) eye out for brownies while I walked around the forest, soaking up ambience.

In Sitka, I was fortunate to meet many gracious people who went beyond the normal bounds of courtesy. To name some: Glen Jones, Joe Miller, Stephen M. Wall, Tim Schwartz, Betty Jo Johns, Burgess Bowder, Carolyn Evans, Bonnie Richardson, Eric Calvin, Ron Bellows, Robert Kite, Alvin Rezek, Ingvald Ask, John Brooks, George Eliason and Paul A. McArthur. Lastly, and most particularly, I offer my most heartfelt gratitude to Lynne Chassin-Kelly, for her good humor and constant support during what was a difficult time in her life.

I also received a warm welcome from the families and friends of the main characters of this book. There were the Doyles, the DeCapuas, the Hanlons, the Morks, the LeFeuvres, the Evanses, the Conners—and not least of all, Robert Carrs and Tamara Morley, who offered insights into the character of the *La Conte*'s skipper, Mark Morley. This is a book about people in a storm, not just a storm; without the help of these kind people, it most certainly would not have held together.

I've always respected the work of the U.S. Coast Guard, but today I say without reservation that I feel a special affection for the men and women who wear the uniform. Not once in four years of research did a Coastie snub me or treat me without the utmost respect. The Coasties I met were generous, accommodating, friendly and professional. I cannot thank them enough.

I owe much to my agent, Owen Laster, for his patience, advocacy and vision, and to Paul McCarthy and his wife, Chicquita, whose belief in the book and suggestions about its structure were vital. I also consider myself lucky to have been able to work with my editor at HarperCollins, Dan Conaway, and his assistant, Jill Schwartzman. Dan's intuitiveness, his understanding of narrative, his line-by-line criticism and the reasons behind that criticism were invaluable.

Finally, to three of my dearest friends, Dolores Barclay, Michael Van Weegen and my mother, Anne: without your unflagging confidence, understanding and encouragement, I might not ever have completed this story-puzzle. With much love, I thank you all.

This is a work of nonfiction—unlike a novel, none of the characters, situations and places described in this book is imaginary. Dialogue was reconstructed as carefully and completely as possible, using official reports, court, papers, personal letters, diaries and audiotapes, and by cross-checking the recollections of people who took part in the conversations. For clarity, consistency and tempo, a number of less significant places, people, observations and impressions have been left out. However, the author has attempted to write an absolutely true book; nothing beyond what was verifiable and documentable has been added.

Nature is always lovely, invincible, glad, whatever is done
and suffered by her creatures.

All scars she heals, whether in rocks or water or sky or hearts.

—JOHN MUIR

AMERICA'S LAST FRONTIER

FAIRBANKS

Denali National Park

A L A S K A

BETHEL

ANCHORAGE

Lake Clark N.P.

Katmai N.P.

Shuyak Island

Bering Sea

KODIAK

Kodiak Island

ALEUTIAN ISLANDS

SHUYAK ISLAND
The northernmost island of the Kodiak Archipelago, Shuyak is home to about 300 of the world's largest brown bears. On August 12, 1998, two teenagers hunting for deer on the island found the remains of a fisherman in a bear den.

KODIAK
This fishing port is the site of one of the largest Coast Guard air stations in the U.S. On the night of January 30, 1998, a snowstorm crippled the base. Nevertheless, the ready crew managed to launch a C-130 plane to assist in the rescue on the Fairweather Grounds.

THE ALEUTIAN LOW
Every winter, this massive low-pressure system develops south and east of the eponymous island chain and tends to spin off hurricane-force storms that track north and east into the gulf.

L

Gulf of Alaska

N.W.T.

CANADA
UNITED STATES

YUKON TERRITORY

Wrangell St. Elias
National Park

Kluane N.P.
Reserve

★ WHITEHORSE

BRITISH

COLUMBIA

LDEZ

YAKUTAT •

Glacier
Bay N.P.

★ JUNEAU

• HOONAH

Fairweather
Grounds

ELFIN
COVE

Graves
Harbor

SITKA •

Baranof
Island

KETCHIKAN •

FAIRWEATHER GROUNDS

About seventy miles from land, these shoals are a favorite among Alaska's fishermen because of the variety and abundance of fish to be found there. But deep-sea currents that upwell at these seamounts make the waters around the grounds turbulent—and oftentimes deadly.

ELFIN COVE

Before making their last run to the Fairweather Grounds on January 29, 1998, the crewmen of the *La Conte* took refuge from heavy seas and high winds in this tiny fishing village near Glacier Bay.

SITKA

On the western side of Baranof Island, Sitka was the *La Conte*'s home port. The night of January 30, 1998, the Coast Guard launched three rescue helicopters from Air Station Sitka in response to a distress signal from a radio beacon on the Fairweather Grounds.

DEEP-SEA GRAVEYARD

100
100
50

The Hambone

The Boot
50

Fairweather Grounds

48
The
Triple
Forties
43
40

The
West
Bank
50

Numbers
represent
depth
below sea
level in
fathoms.

48
46

1000
100

The western shoals of the grounds—The Triple 40s—where the *La Conte* fished for yellow eye and eventually sank on January 30, 1998.

0   100 MI
0   100 KM

BOOK ONE

Not long before finding the dead man, the two boys lowered their rifles and squatted beside bear tracks they could not have imagined. Unimaginable, they were, not because of their width or length—although a couple of hefty loggers could easily have stood together in each paw print—but because the prints were so deep. Whatever had left such indentations in the earth had truly been a creature of great weight and, it would logically follow, of great appetite. The scat piled in high, vaporous mounds along the sides of the tracks only served to confirm it: on Shuyak Island, the northernmost isle of the Kodiak archipelago, there was one humongous—and hungry—Alaska brown bear rummaging about.

Such a discovery might have given other hunters pause. Doug Conner and Jesse Evans gazed in wonder, a flush rising into their faces. The great tracks had stirred in them a curiosity, that careless, youthful desire to glimpse reality beneath the surface of pretty appearances, and they wanted to see more. They were seekers of deer but also of adventure. If they turned back then, what tales could they tell their parents once they had returned to the fishing boat anchored in the cove? No, turning back was not an option. They were on the edge of something, of something harsh, perhaps, but they certainly could not turn tail and run simply because of a few piles of steaming dung.

And so they stood in silence at the edge of the beach, the sweat on their brows shiny in the sun of midmorning, and did what they figured woodsmen normally do before pursuing any creature into unknown territory: they smoked. They puffed on two Marlboros nicked from a pack left carelessly by Doug's father on the galley counter, split a Snickers bar and feigned indifference to the possibilities awaiting them in the forest that stood darkly at the top of the embankment.

Then they stomped out their cigarette butts and let curiosity lead them up the hillside.

It led them along a steep escarpment by the cliffs, to the top of a rimrock formation. They paused and breathed deeply the odors of an aging forest: the dewy tussocks and lichens, the decomposing bark, the crushed, clawed-up tick moss, the mossy, fallen antlers of fawns.

Ahead stretched more trees, more silence, more shadows.

"Listen," Doug said.

"I don't hear anything."

"That's what I mean."

They crept along as noiselessly as they could, the blood drumming in their ears and chests, and halted. Through the smoky light beneath the great spruce they saw the trail wind to a brook and then disappear into thin air.

"Stay behind me," Doug whispered.

"Okay."

"But not too close."

Doug scratched where a bead of sweat had slid down his neck, and began stepping in the prints left by the bear. He carried his rifle loosely, as low as his thigh. The jut of the barrel felt cold in his hand. It was a Remington Redfield .308, a fine deer gun—slim, compact, light trigger. Good for a quick gut shot. Jesse had a .4570 brush rifle with peep sights. It had a short barrel, high calibration. The silence quickened his heartbeat. He thought he ought to rehearse his reaction to a bear charge. He would have enough time to get off two rounds and perhaps Jesse would be able to shoot twice, too. He was not sure if four bullets would be enough to stop an angry sow bounding toward him with the speed of a racehorse, but they would have to do. There were certain moments, his father always said, when a hunter had no choice but to leave his fate in the hands of the Almighty.

Doug looked up. In the high tangle of branches he noticed a nest of eagles. He could see the white crowns and beaks. It occurred to him that they might see something at that very moment that he himself did not. He swung his head around. Nothing. He wiped his trigger hand, cold-drying wet, on his trouser leg. Then, hearing a breath of wind and the piling of surf, he took another step.

And felt it under his boot.

He took a step back, tentatively, and looked down. In the trail, between the scat and paw prints, lay a mitten. A red neoprene mitten. Teeth gouges on the cuff.

"Hey," Doug said.

The other boy came over.

"What you make of this?"

Jesse reached down and picked up the mitten. It felt heavy.

"Cut it open," Doug said.

Jesse snapped open a pearl-handled hunting knife. He slit a cross through the palm, turned the mitten over, shook it. Out spilled sand and . . .

He jumped back.

"That's . . . that's . . ."

Now at his feet in the middle of the trail lay five very soggy, very white bits of skin. Attached to one was a gleaming, human fingernail.

The body came in with no eyes or ears. There was no nose, either. No chin, no teeth—not a single, distinguishing facial mark. There was, in fact, nothing to work with from the neck up, except for a few strands of brittle, soiled hair.

"This is all?"

"That's it."

"Where did the kids find him?"

"In a bear den."

"What?"

"That's right."

"What the hell were they doing in a bear den?"

"Hunting."

"Some kids."

It was chilly in the room. The air-conditioning might have been set too high. The air hung stagnant and smelled faintly of ammonia. Under the hard, fluorescent lamplight, the weathered bits of what once had been a man looked insignificant in repose.

The investigator rubbed the gooseflesh on his arms. "Well," he muttered, "let's get this show going."

"All right," the pathologist said. "Where's that recorder?"

"Here."

The pathologist leaned over the examining table. He wore thick glasses over red-rimmed, blue eyes; a white mask covered his mouth and nose. He had long, bony fingers gloved in latex. He pressed the record button.

For the record he gave his name, Dr. Michael Propst, the date, August 14, 1998, the time, 2:14 P.M. He cleared his throat and began describing what he saw. There was noteworthy biological material. Specifically, bone fragments, hair and skin. The bone fragments showed predation—a large bear, most likely, judging by the depth of the marks. The bones, from an arm, a rib cage, a leg, appeared to be human. So did several of the hairs, although some of it had likely once belonged to a deer or a seal. There were fragments of a neoprene wet suit, also very much predated, among them a sleeve, a right mitten, a trouser leg and a large section of the upper chest. And there was clothing: two white socks, briefs, a T-shirt, sweatpants and a Casio watchband.

Propst coughed.

The box, he went on, also contained a fair amount of dirt, spruce needles and other organic debris from the crime scene. He paused the tape.

"They bagged this stuff in a hurry," Propst said. "Can't say I blame them."

"Anything else?" the investigator asked.

"Yes."

Recording again, Propst made note of five skin fragments. Some had decomposed more than others. One had had a fingernail still attached to it. The nail, he said, was most likely human.

He stopped the tape and pulled down his mask.

"Man didn't clean under his nails."

"No. What else we got?"

"Look here."

Now the investigator, whose name was David Hanson, leaned over the table. On the chest of the suit was an emblem, a penguin, and the brand name IMPERIAL stitched across it.

"The manufacturer," Hanson said.

"Apparently."

Hanson ran his forefinger along the inside of the collar and pulled out a tag. On it was a serial number, a lot number and a date of manufacture: March 23, 1989.

"Not new," he said.

"No."

"And still wet," Hanson said. The material had a loamy smell, like decaying wood chips. "Hang all this stuff in the back room," he said to the lab assistant. "Let it dry out."

As the assistant collected the shreds, Hanson worked his lip with his teeth. He was wearing his navy blue suit, with starched white shirt, tie and black leather shoes. He was neat, clean, shaved and stern. He'd been a cop six years but a member of the Anchorage crime unit just twenty-two days. He was twenty-eight and this was his first case. It would be nice to get it to go somewhere besides a missing persons file cabinet.

He sighed.

"Is this really all we've got?"

"There is the skin," Propst said.

"Right."

Of the five skin fragments the biggest was no larger than a dime. Three others were brittle, yellow, and the last two as sturdy as wet newspaper.

Hanson squinted at them.

"So?"

Propst scratched a fleshy jowl. "Well," he said, "they're human. Remember, they came from inside a survival suit."

"I remember."

"If the fingerprint lab could make prints from these fragments, then you could check the prints against what's on file."

"And you think we could make prints from these little things?"

Propst snapped off the latex gloves. "Ask Walter MacFarlane that."

Walter MacFarlane laughed a smoker's laugh, coughed a smoker's cough and then smiled. A big, Savannah smile.

"You've *got* to be kidding," he drawled.

Hanson smiled back. "No."

MacFarlane's smile soured. He sighed. "All right, what am I supposed to do with this?"

"Make prints," Hanson said. "Or *a* print. Then ID it."

MacFarlane reached for the magnifying glass. "What makes you think these came from fingertips?" He was examining the fragments now.

"The fingernail."

MacFarlane motioned to his apprentice, Dale Bivins. Bivins was short, with dark hair buzzed to the scalp, a prickly complexion and the eager eyes of a freshman on his way to his first college football game. He'd been at the crime lab almost a whole week.

"Say, Dale," MacFarlane said, "why don't you take a look at some real old, real dried-out, real decomposed skin?" He removed his glasses, rubbed his eyes with his palm and took a step back.

"David," he said, "you're asking me to ID a guy who's got enough skin left over to fill half a matchbox. Normally, that might be tricky. In this case, it'd be like winning the lottery. Look at this skin. As soon as we touch it, it's going to crumble like a cheap cookie."

Propst said, "Walter?"

"Yeah?"

"Did you see this fragment?"

"Which one?"

"This one."

MacFarlane put his glasses back on and squinted at the specimen on the table. "I did."

"Could you work with it?"

MacFarlane paused. He was trying to figure out if Propst was kidding. No, obviously not. MacFarlane shook his head.

"All right," he said. "Send it on over. I'll give it a shot. But this one here's a million-to-one horse."

Hanson nodded.

"No promises, now," MacFarlane said.

"No promises," said Hanson.

T W O

As David Hanson saw it, identifying his mystery man hinged on two leads: the survival suit and the skin. The suit was the more promising. All he had to do was track down the original owner and work forward. True, the suit was nine years old. It could have changed hands a few times. But someone had to know who'd been using it before it washed up on Shuyak.

It occurred to him that there ought to be a registry of survival suits somewhere. The Coast Guard kept records on almost every piece of equipment on commercial fishing vessels, even something as obscure as an emergency satellite beacon. So why not a registry of survival suits?

As far as the skin, he thought, Walter MacFarlane was probably right. It was a lot to expect a clean print to be made from so tatty a fragment. Even if the lab could do it, what were the chances of getting one print to exactly match another in the FBI's national database? There had to be more than 35 million sets of fingerprints in that database.

All right, Hanson thought, it's a long shot. So what? This poor guy had a life and family and right now, somewhere, that family is probably still leaving the front door open for him.

Maybe I can do something for those people, he said to himself. Help them close that door.

His first call was to the Coast Guard air station in Kodiak. He asked for the Marine Safety Office. That was the clearinghouse of maritime records. A clerk answered. Hanson told him who he was and that he wanted a serial number from a survival suit.

"I'm sorry, Mr. Hanson. That won't be possible."

"Why not?"

"We don't keep those records."

"Well, who does?"

The Imperial Company was headquartered in Philadelphia but had an office in Bremmerton, Washington. It took Hanson close to a half hour to get a correct number out of directory assistance, and when he did get one and dial it, the call ricocheted around a half-dozen departments. Finally he heard a woman say, "Carey Guddal!" in a bright, chipper voice.

He went through his introduction.

"I've even got a suit number," Hanson said. "It's 109153—"

"I'm sorry. Mr. Hanson?"

"Yes?"

"I'm afraid I can't help you." The voice had not lost its cheerfulness. "Our parent company made that suit, and it's no longer in business."

"Right."

Next he tried the Kodiak state troopers. He asked for a list of people who had gone missing on or around the Kodiak archipelago for the past decade. He also remembered to ask the sergeant to put an ad in the local paper asking the public for tips on overboard boaters.

Afterward, he visited the forensics lab and requested a DNA analysis on the hair, skin and bones collected on Shuyak. And then he went to lunch.

Just before five that afternoon his telephone rang. It was Carey Guddal again, with her chipper voice. The more he listened to her the more he slumped in his chair, until, finally, he just thanked her and hung up. He sat there, thinking, and then straightened up and clicked on his desktop computer. He typed the date and time at the top of a blank page and wrote:

> *Guddal recontacted me later in the day and informed me that Imperial was purchased several years ago by an East Coast real-estate developer named Michael Callahan. Callahan quickly sold the company. After the company was sold, most of the records concerning the individual survival suits were either destroyed or filed away in a warehouse somewhere. According to Guddal, all of the people who used to work for the company have since taken a new job. She did not have any other contact information for any other field employees.*

Hanson read the words over, saved the file and cleared the screen. He turned off the monitor, leaned back in his chair and looked out the window. Then he went to the coatrack and picked off his jacket. Not bad, he thought. You killed your best lead on the very first day of investigating. Maybe Walter MacFarlane will be luckier with the fingertip skin.

Four of the tissue specimens were useless. They were too small, too split, too dried out, too curly.

The fifth tissue sample, which had come from a right forefinger, wasn't bad. It had the circumference of a dime and the ridgelines on it were decent. But it was extremely fragile. Stretching the skin even a fraction of a millimeter would cause the ridgelines to distort. That would make any print reproduced from the fragment useless.

So they let it stew.

More precisely, they left it in a petri dish filled with a solution of liquid formaldehyde and fabric softener. The softener was supposed to loosen up the edges that were brittle. The formaldehyde, with luck, would dehydrate the soggy middle and firm it up.

After lunch, a lab assistant plucked the tissue out of the solution, sandwiched it between two glass slides and delivered it to the fingerprint lab. Walter MacFarlane studied the skin under a high-powered lens.

"It's still okay," he said. He stuck a cigarette between his lips but did not light it. "How about I dip this tissue in ink, pick it up with my forefinger here and roll a print out on the tracing paper?"

"That won't work," Dale Bivins, his apprentice, said. "It will smudge. It will come out blurry. The tissue is too fragile, besides. We should not press it."

"Well," MacFarlane said, with a thin sigh, "I'm fresh out of bright ideas."

Bivins sat up.

"How about a clay finger?"

"A what?"

"A clay finger."

MacFarlane gave Bivins a blank stare.

"We make a clay hand," the apprentice said. "With a forefinger. The forefinger points up and out, see, and then we drape the skin on the tip of the finger and then shoot a photo of the ridge detail with the bellows camera."

MacFarlane's eyebrows went up as he listened.

"Then we enlarge the photo."

"That," MacFarlane said, "is one of the craziest damned things I ever heard." He smiled. "I love it."

From a lump of clay Bivins sculpted a fist. Then he added a forefinger. Then, with padded microtweezers, he gently hung the skin fragment from the tip of the finger. He took the clay hand to the camera room and positioned it in front of the Crown Graphic, an inch from the lens.

"You know," Bivins said, "the tissue is split in two places. I'm not sure about this."

"Hell with the splits," MacFarlane said.

Within an hour, Bivins had a negative. He produced a large print from it, dried the print under a heat lamp and then placed a sheet of tracing paper over the photograph and traced the ridge patterns onto the paper. Then he took a picture of the patterns he had traced. And then he developed the negative to a one-inch-by-one-inch size.

On the fluorescent viewing panel the print looked all right. The loops looped, the swirls swirled and the ridgelines were sharp and true.

"That," MacFarlane said, "is damned nice."

They scanned the image into the Automated Fingerprint Identification System and waited. Twenty-four minutes later, the printer spit out a list. There were eight names on it. They pulled all eight sets of fingerprints from the archives and brought them back to the lab.

MacFarlane examined the first card and set it aside.

"Forget him," he said.

"How come?"

"This nice fellow is still in jail."

The next five sets of prints came from people who were still living. The last two ten-finger cards had prints that were close, but not exact matches.

Bivins shook his head.

"What now?"

"Now I go out for a smoke," MacFarlane said.

The weekend slid by. The first thing David Hanson did when he arrived at work the following Monday was to head straight to the medical examiner's lab. No one had come in yet. The door was open, so he entered, walked to the back and stopped before a door. He tried the knob. Unlocked.

He found the switch and a bulb flickered and then lit up a ceiling fixture full of dead flies. The tatters of the survival suit were still hanging from a line. They smelled like boiling alcohol under a blanket.

Reaching out, he touched one of the strips, probably once part of the suit leg. It felt dry, scaly. He removed his coat, rolled up his sleeves. He tore a sheet of butcher paper from a wall dispenser, spread it out over an examining table, took down the chest portion of the suit and laid it on the paper. It smelled. He looked askance at it. Then he wiped his hands on his slacks.

Dark in here, he thought.

He went to the window and pulled a cord. The blinds went up and sunlight burst into the room, powerful and perfect, like a flash of lightning at noon.

Again he considered the suit.

His eyes combed the material. The gouges meant nothing to him now. He turned it over. Across the back was a strap. On it were three, faint letters, scrawled crookedly in black marker:

B O Y

He stiffened.

Why, he wondered, didn't I notice that before? Wait a minute. The material was damp. It was damp and the orange color you're seeing now was *maroon*. That's right. Those black letters must have blended in with the maroon color. That's why you didn't notice them.

Leaning closer, he picked out another letter:

M B O Y

His pulse jumped in his arms.

He thought: Is this a code? A nickname? What? A ship name? He stood there, gazing at the strap. As he did, a fifth letter appeared on the fabric, as though he were seeing it take shape through a fog.

And then, a sixth letter.

T O M B O Y

————

The room was long and narrow with gray walls, no windows and a door that opened to a basement hall. The linoleum floor glared under the fluorescent light. There was an unreal, bluish-white quality to it, like light filtered through an aquarium tank. Black file cabinets stood in a row in the middle of the room. These were the John and Jane Doe cabinets. They were stuffed with files of missing, forgotten people, and rarely ever opened, except to add new folders.

On one wall hung a map of Alaska stuck full of colored pins. The red pins designated where people had gone missing aboard ships. The blue pins showed where passenger planes had vanished. Green pins marked where mountain hikers had disappeared and the yellow ones indicated where folks with a good reason to vanish had vanished. There were many black pins, too. These marked places where unidentified bodies or body parts had turned up.

David Hanson sat counting the pins. Across from him sat a short man with a potbelly, oily hair and a nose with veins that glistened in the harsh light. He had tight, watery eyes behind bug-eyed glasses and the slow, deliberate movements of a night watchman.

"So you got yourself another lead?"

"I think so."

Sergeant David Johnson, director of missing persons, leaned back in his chair and clucked his tongue. He said, between clucks: *"Tomboy?"*

"Uh-huh."

"You think it's a nickname?"

"Could be."

"A boat, maybe?"

"Maybe."

"Fish and Game," said Johnson. He yawned, turned ponderously in his swivel chair, pulled a computer keyboard out from under his desk and started pecking at the keys with two fat, neat forefingers.

"What are you doing?"

"Searching the Fish and Game database," Johnson said. "You want to find out who this *Tomboy* fellow is, right?"

"Yeah?"

Johnson kept pecking at the keyboard and drawing air through his crimped, whistling nose. "Let's see. Fishing vessels . . . Alaska . . . registries . . . Okay. Right. Here we go." He peck-peck-pecked, moved the mouse, clicked, peck-pecked, and finally, with a satisfied cock of his double chin, said:

"Three."

"Three what?"

"Three *Tomboy* vessels."

"Alaska registry?"

"Yep."

The first vessel, registered in 1989, had been sold the following year to a Fred Tomkoff Jr., the current owner. The second vessel had been bought and registered in 1989 by a George A. Shapley Jr. The third boat had been purchased eight months earlier, in January of 1998. It was co-owned by an Arthur Eels and a Daniel W. Minor, both of Port Alexander.

"How about a printed copy?" Hanson said.

"Of course." Johnson handed him a sheet of paper. "Here. This ought to narrow your search."

"Hopefully."

"Well"—Johnson offered his hand—"if there's anything else you need."

"Thanks."

Hanson stood and looked once more at the big map. He noticed a black pin on the northern tip of the Kodiak archipelago. Johnson stood beside him and raised his eyebrows, as if raising them was an effort.

"I hope you find out who this *Tomboy* guy is," Johnson said. "We could do without the extra work around here."

There was no record of any fishermen with a first, last or middle name of Tomboy in the Alaska Personal Information Network database. So Hanson went through his mail pile. There were some letters, an official one and some others. The official one came from the chief of the state troopers in Kodiak. It was the missing persons list Hanson had requested. He opened the manila envelope. Between 1980 and 1992, fifteen people had gone missing in Kodiak. After 1992, the troopers had stopped keeping a tally, but the report did not explain why.

It was unlikely that skin tissue would last six years in a survival suit on a bear-inhabited island. Hanson ran the names through the computer anyway. Of the fifteen names on the list, the prints of six were still on file. He asked Walter MacFarlane to see if any of them matched *Tomboy*'s print. They had all taken to calling their mystery man *Tomboy*.

That was the last new lead Hanson would get for eight days. By the afternoon of August 24, the investigation had stalled again.

According to the Coast Guard, no *Tomboy* vessels had ever sunk in Alaska. According to forensics, there was no chance of a DNA analysis on the hair, skin and bone fragments. The skin and hair were—how did they put it?—*unsuitable* for testing. Too decomposed. The bones could be ground into a powder, tested in a beaker, but without a family donor to offer a comparative sample, what good could it do?

The state's Fishing Entry Commission had issued fishing permits to the *Tomboy* vessels but had no other information. The fingerprint lab could not match the *Tomboy* print to any of the six people who had gone missing on Kodiak between 1980 and 1992. And calls to the National Oceanic and Atmospheric Administration, the Washington State Department of Vessel Licensing and the Coast Guard's National Documentation Center in West Virginia got Hanson nowhere.

For eight days he left phone messages for the men who owned *Tomboy* vessels in Alaska. He got only one callback, on August 24, when George A. Shapley Jr. phoned to say that his *Tomboy* was still afloat, and all crew and survival suits accounted for. "That's great news," Hanson had told him. "Thanks so much for calling."

The following afternoon, however, a phone call from the Kodiak state troopers perked him up.

A team of investigators had gone back to Shuyak Island and found more human remains. The discovery was not two hundred yards from the bear den. Everything was being dispatched to Anchorage in the morning.

It was the talk of the crime lab until two the following afternoon. Everyone assigned to the case gathered in the medical examiner's lab. But when Dr. Propst opened the sealed carton, their jaws dropped. There was no more than the sole of a shoe, a pair of tattered, gray sweatpants, the sleeve of a sweatshirt, two socks, long johns and a few bone chips.

The next morning, Hanson found a message on his telephone answering machine left by someone named Geraldine Dodge in Kodiak. She'd also left a return phone number. He dialed it not knowing quite what to expect, and heard a woman's voice.

"Ms. Dodge, this is David Hanson from the state crime lab in Anchorage. You called me?"

"I did," she said. "I've got something that might help you solve that missing persons case on Shuyak."

"Be my guest."

Dodge told him that in 1997, a fisherman by the name of Thomas A. Banks had disappeared in the Gulf of Alaska, eighty miles south of Cordova. He was on the *Cape Chacon*, working the deck, when a wave swept him overboard. The Coast Guard had never found him.

"What makes you think he washed up on Shuyak?" Hanson asked.

"I don't know—maybe his nickname?" Dodge said. "Tomboy."

For some reason, Banks had never been registered as a missing person. The Coast Guard did have a case file on the mission, however, at the Marine Safety Office in Juneau. Dodge provided a telephone number.

The ensign was almost apologetic. He could not recall any search-and-rescue missions involving a Thomas A. Banks or a fishing vessel *Cape Chacon*. The only boat to go down in the Gulf in recent memory, he said, was an old trap tender called the *La Conte*. It had sunk on the Fairweather Grounds, seventy-five miles southwest of Yakutat, in January.

"Yakutat?"

"Yes, sir."

Yakutat was a village on the eastern rim of the Gulf of Alaska, north of Glacier Bay. That had to be more than seven hundred miles away from Shuyak, Hanson thought. There was no way a body would float that far without a fishing boat spotting it or pulling it up.

"Ensign?"

"Sir?"

"Any chance you could run a check anyway on that *Cape Chacon* vessel?"

As it turned out, the *Cape Chacon* did lose a crewman once, only not in 1997. The fisherman, Thomas A. Banks, had gone overboard near Cordova in 1987—two years before the *Tomboy* suit was manufactured.

Hanson ran the name through the state computer anyway. Banks had prints on file, so Hanson asked the fingerprint lab to compare them against the *Tomboy* print. The results came back after lunch.

Negative.

David Hanson got up the next day feeling rested but sluggish. He showered, shaved, put on slacks, a clean white shirt, his black shoes and a bright tie and looked in the mirror. He undid the tie and walked down to the kitchen. He cleared a spot on the table, sat down, had a few handfuls of his kids' cold cereal, a shot of cranberry juice and thought about *Tomboy*. The kids were loud and his wife said something to him but he did not hear it. He felt drugged, sort of. He put his coat on and went out to his car.

Driving, he kept thinking about the survival suit. The windshield wipers squeaked back and forth. What had he missed? He felt as though he had overlooked something. There was always something. Life was a series of clues overlooked. Who had said that? He pulled into the parking lot.

Three o'clock found him at his desk, typing. He had kept a running account of the *Tomboy* case and did not want to stop just because the investigation had flamed out. As he wrote, it occurred to him how many concrete leads had gone up like a puff of dust in a draft, and he kept on writing. He made a note that he never had contacted two of the men listed as owners of *Tomboy* vessels in Alaska. He'd called several times and left voice messages. Perhaps he ought to try them one more time. He reached into his drawer, took out his notebook and was flipping through the pages to find the numbers when the telephone rang.

It rang a second time.

He put his hand on the receiver, lifted it and said, "Hanson."

"Yeah," a voice on the line said. It had a thick edge to it. A three, double-Scotch edge. "This here's Eels, from Port Alexander."

"Eels?"

"Yeah, some guy named Hanson called me and I'm returning the favor."

"Arthur Eels?"

"Yeah."

I was just going to call you, Hanson thought. What he said was, "Arthur, I'm glad you called."

Silence.

"Now, Mr. Eels—"

"Arthur."

"Arthur," Hanson said. "Listen; the reason I called was because you're listed in Fish and Game records as being the owner of a boat called the *Tomboy*—"

"I sold that boat."

"When?"

"A couple months ago."

"I see. Well, who did you sell it to?"

Eels gave him a name, then said, "I got myself a new one. The *Emily Ann*."

"So you no longer own a *Tomboy* fishing boat?"

"Not anymore."

Well, Hanson said to himself. Another strikeout. "Okay, well, Arthur, I, uh—I appreciate you calling back. You told me all I need to know. That is, unless you think there's something else you can tell me."

"I don't think   ''

"Right. Well, thanks again."

"I mean," Eels went on, as if he had not been listening, "you probably already knew that one of the guys who died on the *La Conte* was wearing one of my *Tomboy* survival suits."

Hanson sat back in his chair.

"Oh, *really*?"

"You didn't know that?"

"No."

Eels, who fished commercially during the summer months, told him he had taken a vacation to Oregon in December 1997, and had left his boat, the *Tomboy*, in his brother's care. His brother, however, had gotten arrested and asked one of his buddies to keep an eye on it, a friend by the name of Mike DeCapua. Shortly thereafter, the friend—Hanson was scribbling down the name—had taken a deckhand's job on an old schooner, the *La Conte*, which was short one survival suit.

"And wouldn't you know it?" Eels said. "But that old boat went down in a storm last January on the Fairweather Grounds. And get this: the guy who died in that storm was wearing my suit."

"Imagine that," Hanson said.

That conversation David Hanson recorded carefully in his ledger. He noted the time and date it took place: 3:37 P.M., Friday, August 28, 1998.

The following Monday, the Coast Guard's District 17 headquarters in Juneau confirmed that two of five crewmen on an old schooner named the *La Conte* had died in a violent storm on the Fairweather Grounds on January 30, 1998. One of the dead fishermen had never been found.

Hanson jotted down the name of the missing fisherman. He ran it through the state's personal information database and found that the dead fisherman had fingerprints rolled on May 11, 1968. Those prints remained on file at the Alaska state crime lab in Anchorage.

Walter MacFarlane pulled those prints. He put the right forefinger on the lab's FX8B Forensic Optical Comparator, side by side with the negative of the *Tomboy* fingerprint.

He and two other experts found them to be identical.

David Hanson was at his desk, typing, when a knock came at the door of the investigator's office. His fingers stopped and he looked up. The door was open. Walter MacFarlane was standing in it.

"Well?"

MacFarlane grinned.

"It's him all right," he said.

"That's great," Hanson said. "That's really great, Walter."

"I can't believe it," MacFarlane said. He held out a fingerprint card with a big circle around one of the rolled ink impressions. Hanson took it. "I'll put it all in the report and get it to you by tomorrow afternoon. It's pretty amazing."

"Yes, it is."

"Congratulations, David."

"No," Hanson said. "You did a great job."

"It's a million-to-one come in," MacFarlane said. "Never happen again, by God."

"You did a great job, Walter."

"No, no, David. You did."

After MacFarlane left, Hanson leaned back in his chair and looked blankly at the fingerprint card. Walter was right. It *had* been pretty amazing when you thought about it. It was as though they had been *meant* to identify this guy. How many pins are there on that big map in Missing Persons? Must be hundreds. Hundreds. So why is it that we got this one?

He straightened his tie. There were in-house notifications to make. He would start with his sergeant. I can't wait to see Sergeant Marrs's face when I give him the news, he thought. I'll bet he's already written this case off. Hanson stood up. He studied the fingerprint card again. It is sort of weird, though. All we had was a dime-size bit of skin and a strap on a suit that read *Tomboy*. Well, forget it, he said to himself. Just leave it be. It'll just tie you up in knots. That's what happens when you ask too many questions. You got to know when to stop.

BOOK TWO

A year earlier, in November of 1997, the first very cold nights came early, then the afternoons were cold and the lamps along the docks began to come on early and Bob Doyle knew the fall was really gone. The salmon were no longer running up Indian River and the black bears that had been a nuisance in town all summer had gone back up into the mountains. The mountains had changed, too. All summer they had been a golden, spruce green except for the very tops that were always white, but now the snowline was coming down a little each day. One day in the middle of the fall, as Bob Doyle walked near the Old Russian Cemetery, he noticed the line was markedly lower on the mountains and he knew that winter would soon be upon all of them.

More boats were coming in now than going out and many were going south for the winter. The fishing along the outer coast was over and by the middle of the month the shrimping season was about over, too. The last of the cruise ships had shoved off to Seattle and the souvenir shops with the not-so-cheap ornaments around the cathedral and on Lincoln Street across from the harbor had sales signs in their display windows. In the harbors there were a few big trollers. But their gear had been stowed and their galleys and cabins stripped and the boats rocked sadly in their slips, the tide licking their hulls and making the bowlines creak, the wind moving their wire stays with a hollow tinkle. All of the noises of the docks now blended in a single note, a hollow note, as though a piano player had hit his last key but had kept a foot on the floor pedal.

That fall the rains came almost every day. The clouds would brood over the mountaintops and then the sky would descend, heavily, and then everything was gray and the mountains gone and the rain black and lisping along the pier. Some days a fog covered the town and Bob Doyle would put down the shrimp pots he was repairing along the wharf and watch the O'Connell Bridge dissolve in it. He knew the bridge was still there but sometimes he liked to pretend that the fog had wiped it away, erased it, and that later, during the night, it had somehow been rebuilt from scratch. Other days the wind blew very hard and the silvery sky would lift and brighten and, pulling apart, allow tilted shafts of sunlight to fall through, lighting a rainbow. The rainbow would not last long. The sky would anger to black and soon the rain was coming again in gray, sweeping nets, lifting the channel in white, spurting jets and dripping from the tails of the ravens perched on the pilings.

All of the sadness of the town came with the cold rains and there were days when he could not see the snowline, only the dripping streetlamps and the slick grayness of the sidewalks and the moldering roofs and shutters of the older cabins. It was not the most pleasant weather for walking but it was easier for Bob Doyle to think clearly when he was out roaming. He also found it more economical to walk than to drink and he had no other means of getting around. So he walked. Only sometimes as he walked would he feel as though someone was behind him, stepping in his footsteps. He would hear rustlings and voices worn away by the years. And sobbing. He would hear it for a time, mixed with the sound of the rain. When he heard the sobbing he had to tell himself not to turn around. Maybe a day would come when the echoes would die.

The best places to wander in Sitka were on the main street and along the wharves. There were many ways to get from Georgia Kite's, where he paid to sleep on a couch, to the waterfront. The quickest was down Jeff Davis Street. He would walk in the opposite direction of the veterans' cemetery, make a right at the tennis court and then a stroll along Lincoln past Crescent Harbor. He had never seen anyone play on the court and he often wondered, in a place that saw more than 270 days of rain a year, if anyone ever got a chance to use it. Soon he would see the stone figures of the seals, glistening and lazy looking, and behind them, the harbor. The bigger boats—the *nicer* boats—tied up at Crescent. Many had fiberglass hulls and fresh paint and cabins and galleys as cozy as the rooms at the Super 8 motel over on Harbor Drive. It was a pretty harbor, a picturesque harbor, with the mountains, piney dark and coldly whitecapped in the backdrop, dominating the sound but far enough away to not leave any shadows, and yet he did not like it as much as the other harbors, the ANB and Old

Thompsen. Perhaps it was too pretty a harbor; it lacked an edge. Still, he liked to walk past it on windy days when the boats pitched a little in their slips, their masts wagging this way and that in a strange unison, like the quills on the back of a moving porcupine. Some days he would spot bald eagles huddled darkly atop the masts and he and the birds would stare at one another. They were looking for their next meal, he would say to himself, and so was he. When he could not find work he might take a bench on the concrete landing and watch the eagles and listen to the surf under the pier. It was a lonely sound but a soothing sound as well, and after a while of sitting in a trance he would snap upright and glance around, as if a hand had shaken him a moment after he had nodded off.

Since he usually had no appointments, no invitations anywhere and no money in his pocket, he often stood alone on the O'Connell Bridge and sipped from a can of warm beer. It was always pleasant crossing bridges. Sometimes he thought he should have lived somewhere like Paris. Sitka had only the one bridge. It linked Baranof Island to Alice Island. Alice Island had the city airport, the Coast Guard air station, the officers' club, the exchange and Coast Guard housing. He had lived on that side of the bridge once, on Lifesaver Street. That was when he had the package: the wife, the kids, the house, the rank, the uniform, the minivan. That was some package, all right.

Now he lived on the other side of the bridge, at Georgia Kite's. The house of misfits. That's what it was, all right. He was a misfit, too. What else, he thought, do you call a guy who scuttles about the docks like a roach looking for broken pots to fix? A guy who bums cigarettes from strangers? A guy who gazes like a human vegetable at the broken reflections of boat rigging in the water, who stops at the same monuments around town to read, over and over, the same old inscriptions?

There was one inscription he especially liked, though. It was the one on the Tlingit canoe out front of the Centennial Hall. *Te Kot Keh Yao*, it read. Everybody's Canoe. Not ten feet away sat the statue of Aleksandr Andreevich Baranov. Baranof. Now, *there* was a misfit. He looked all stern and spiffy, old Baranof did, in his top boots, colonial coat and necktie, sitting there with a scroll in his fist and his gaze fixed on his colony. Russian America. Some colony. Maybe that's why the artist had left a couple of holes where the eyes should have been. There was no looking into the future.

When he tired of the bridge, Bob Doyle liked wandering downtown. There was the Moose Lodge, the Columbia bar, an electronics shop, a pizza joint, a pelt shop, a trinket shop, the Veterans' Hall, Old Harbor Books, the Sitka Hotel, Ernie's Saloon, a pharmacy and, of course, squatting in an oval in the center of the street, the Cathedral of St. Michael. The original Russian church had burned down in the 1960s and the city had built a fireproof replica. A fine replica, he remembered thinking, the first time he saw it. So neat and perfect. Now it bored him. He never did think much of churches. Perhaps it was because he was such a crummy Catholic, but probably not, and he routinely shuffled by the cathedral, hulking gray and white-trimmed and clean looking, but not very inviting and altogether too perfect, on his way to someplace more important, say, the Moose Lodge or the Columbia.

Sometimes he would pause outside of the bookshop. It looked impressive enough through the big windows, the books carefully stacked and arranged, their covers shiny in the display lights. He would read all the covers and on days when the rain was coming down especially hard he would spot a notice scribbled on the side of a grocery bag: PLEASE DON'T DRIP ON THE BOOKS. He rarely entered. He liked standing out on the sidewalk, looking in. Sometimes he might see a reflection of himself in the glass. He would see a tall man, stooped, somewhat below the threshold of slender, with a drawn, unnaturally white face that could have been borrowed from a leaden mask. On top of the man's head sat a rumpled, olive green cap. When he removed it, he looked a lot older. Most of his scalp was bare, shiny at his north pole, with cantilevers red as shrimp tumbling down the back and sides. He had a sharp nose, a long chin covered in a shaggy tuft of red-and-white hair and sad, sheepish eyes. The eyes were devoid of any traces of confidence. Sometimes as he gazed at the face in the glass he would nod, as though nodding at a stranger.

Katlian Street, the road that snaked along the waterfront, he found pleasantly run-down. It had once been all Tlingit clan houses. Now it was the guts of Sitka's fishing industry. Along the waterfront was the Halibut Hole coffee shop, the ANB Harbor, the Alaska Native Brotherhood Hall and the cold storages and supply outlets. There was also the Pioneer Home for retired fishermen, the Pioneer Bar for active fishermen, the Tlingit cultural house, a shop where Natives sold handwoven baskets, painted figurines and beaded crafts, and a soup-and-sandwich place decorated with maps and fishing gear. The rest of the street was sagging frame houses and lots empty except for weeds, bald tires, rusting fuel drums, discarded nets, tarps, buckets, oil cans, newspapers and pallets.

Sometimes when he passed the gift shop Bob Doyle paused to read the faded sign hanging on the glass front door:

> Open most days about 9 or 10, occasionally as early as 7, but some days as late as 12 or 1. We close about 5:30 or 6, occasionally about 4 or 5, but sometimes as late as 11 or 12. Some days or afternoons we aren't here at all, and lately I've been here just about all the time, except when I'm somewhere else, but I should be here then, too.

Below it hung a second sign:

> Sorry, we're closed.

Mornings he would make the rounds looking for work: the harbormaster's office, the pier, the cold storages, the fish brokerages, the packing houses, the supply outlets, then stroll along the waterfront and watch the many gulls and ravens that took off flying when he crept close to them. Afterward he might, or might not, stop to see a friend, Eric Calvin, who kept his small, steel shrimping boat tied up on the quay. Calvin had tried to help him once by bringing him along on a shrimping trip near Tenakee Springs. For almost eight weeks they had worked the inlets around Tenakee, but in the end they came away with barely enough to pay for the Old Milwaukees they emptied at a café there run by a Philippine woman. Calvin had written it off as bad luck. Sure, he had told Calvin. Story of my goddamned life.

On his way home he would stop at the Sitka Job Center on Lake Street. He might toss down a few cups of free coffee, sneak a couple of cookies and scan the bulletin boards. If he saw a job posting, he'd print out a copy of his résumé on a sheet of paper, lick the envelope and hand it to Bonnie

Richards, the manager. He couldn't see why he bothered. The Coast Guard wouldn't give him a good word if he applied for a ticket taker's job at the circus. But Mrs. Richards would smile that terribly bright smile of hers, and he would put on his best face until he felt it pulling out of shape and then he would be out on the street again, the night falling cold and damp and the display windows of the shops already dark. Mrs. Richards. She actually cared about him. Why? He couldn't understand how. He really couldn't. He needed to start doing that—caring about himself. He needed to get some forward motion. Sure. That and some courage.

F I V E

Georgia Kite's house was a picture of decay. On sunny days the outside shingles had the hue of mildew; on rainy days, cigar ash. The roof, drooped in the middle, was covered in chinks of moss as thick as any on a dying spruce. The driveway looked as though it served as the neighborhood dump: stove parts, coffee cans, tarps, sawhorses, snowmobiles, motorbikes, tires, monkey wrenches, a bathtub, grease guns, a pressure cooker, hoses, fuel drums, rolls of fiberglass insulation and a half-dozen three-wheelers in varying stages of dismantlement. The front door was broken—nailed shut—and there was no way to get to it in any case, since the front stairway and stoop were missing. To enter Georgia Kite's, one needed to trump around back, climb the steps of a flimsy, molding stairway and tap on the kitchen door.

The kitchen itself was a shambles, the sink somewhere under a jumble of encrusted pots, liquor bottles and dishes, the floor occupied by stacks of beer cans and tinned food, from Del Monte green beans to Campbell's cream-of-mushroom soup to Carnation fat-free powdered milk. The living room had three couches, a recliner, a coffee table, two TVs, dual VCRs and a pair of radios that usually bleated discordantly on different programs at the same time. On the paneled walls, at peculiar angles, hung a black Eskimo mask, a painting of sailboats on a pink ocean, a feathered, Indian headband, the spoked wheel of a troller and the American flag, upside down and fraying at the edges. Bits of underwear, ashtrays and empty cans lay scattered among the pillows and cushions, and the window shades were stuck to the panes with packing tape. The house had an odd blend of odors, the heaviest of which was the sticky, stale smell of old nicotine, no telephone, no clocks that kept accurate time, vases without plants, leaking pipes that left dark blotches on the tiled ceiling, and carpets sticky with beer and bodily fluids.

There was an upstairs bedroom, which Georgia had fixed up with a queen bed, a lamp she got at a garage sale and a cracked window shade and rented out at $350 a month to a newcomer named Dale. But the real action was in the basement. It had a bar, complete with kegs, blenders, a stash of Stoli, Jim Beam, Canadian Hunter, shotglasses and mixers, bottle openers and a neon Budweiser lamp. There was a Foosball table, two regulation-size pool tables, an electronic dartboard and a poker table. Smiling from the walls were centerfolds in their birthday suits, a poster of Mount St. Helen's erupting and a notice: PLEASE DO NOT FEED THE HUMANS. The couches showed their stuffing, but were comfy, and through a door was a room with a queen-size bed—that kind of room— and a ceiling light that, not so remarkably, never worked.

Some of Sitka's wildest and longest-running parties took place down there, and after a while people around town just started calling Georgia Kite's place the Basement. Once the P-Bar, Pilothouse, Ernie's, the Columbia and Rookies shoved their patrons out at two o'clock, in accordance with the local liquor laws, the rank and file simply filed on over to Jeff Davis Street and carried on in the Basement, oftentimes through the following afternoon.

Georgia didn't mind crowds. She was extra soft on fishermen—her second husband, a cook, had gone down with a fishing boat—but she took in loggers, miners, dope peddlers, faith healers, hypnotists, cheating spouses, college tramps, circus workers, swindlers, grifters, hookers. She took in anyone who didn't molest children or kill people, though, in fact, more than a few felons had found temporary refuge there at one time or another, without her knowing their shady deeds. Regulars got charged a "maintenance fee" of seven dollars a night for a couch, payable upon departure and IOUs negotiable, but visitors were allowed to sleep it off on the floor, free of charge.

Georgia Kite's might have been classified as a roadhouse, if anything like a road network existed through southeast Alaska. Instead, it came to be known as a 24/7 "rest house"—a stop at the end of the line for America's end-of-the-roaders. A landing pad for drifters, dreamers and derelicts, for those who rejected—and got rejected by—suburbia. Not a house for the homeless; rather, it was a tent with a roof and walls—a tent for flower children and forever children, for the professional gone hobo, for rootless men and women who had nothing but each other.

It was the last door in Sitka open to Bob Doyle.

After Bob Doyle's "retirement" from the Coast Guard, the officer who took over his duties in supply had been nice enough to rent him an apartment he owned in town. He was booted after ten days for not coming up with the rent. Later, he rented a trailer out on Halibut Point Road but the same thing happened. He went door-to-door, staying with friends until their friendliness ran out, and then one day at the Pioneer Home, the retirement home for fishermen, an old-timer told him to look up an old gal with a heart as big as Mother Teresa's, Georgia Kite.

He had been untangling longline gear and repairing shrimp pots down at the ANB Harbor, but he was not making anywhere near enough to drink and pay real rent. So one afternoon, after he had woken up with a whiskey rug in his mouth, he packed his old dress blues and some personal effects in two cardboard boxes and turned up at the kitchen door of the house on Jeff Davis Street.

A dog yelped. He squinted through a film of grease on the glass-paned door. A man with dark eyebrows closed in together and a cigarette stuck to his lower lip opened the door. His lean face wore a scowl.

"What?"

Bob Doyle introduced himself, raising a furrow of mild disgust on the man's forehead.

"So you're the Coastie?"

"Ex," Bob Doyle said.

The man nodded. His close-set, dark eyes studied him now without rancor. "Right. Well, I'm Gino. John Gino. C'mon in. Leave your shit wherever you can find some floor."

He left the boxes out on the stoop and stepped inside.

"Want a drink?"

"What you got?"

"Popov."

The man named Gino poured out two shots and they downed them. The door rattled open.

"Hey, Mike," Gino muttered.

The man who had just entered grumbled, "Hey, yourself. Got a cigarette?"

"Nope. Just pinched this one from Dick."

"Who's this?"

"Coastie Bob," Gino said.

"Coastie?"

"That's right. Bob, here, was even an officer once. Now he wants to be a fisherman."

"Really?"

"Ask him."

This second man was early forties, Bob Doyle thought, about six feet in height and wiry, like a scarecrow made of cables. He wore sweats, shin-high rubber boots, a powder blue, checkered shirt and sweatpants that appeared not to have seen the inside of a washing machine in weeks. Rubber bands held his wild, greasy mane together in a ponytail that flopped halfway down his back. He had a scar across the bridge of his nose, brown, broken teeth and a mouth that sliced downward at the corners as though he had just frowned, or was about to. But his mouth told very little about him. His eyes were what you needed to watch. They were blue as a glacier and fierce as the eyes on a junkyard dog.

Gino said, "This here's Mike DeCapua."

"Hi, my name's Bob."

The man considered Bob Doyle's outstretched hand without any haste. "Coast Guard, huh?"

"Yeah."

"Fucking worthless."

Bob Doyle smiled and said, "Okay."

The man named Mike DeCapua was still sizing him up, as he did with all newcomers. His eyes probed Bob Doyle impersonally. "So you want to be a fisherman, eh?"

"Yeah, I guess."

"Well, there aren't many good ones left out there."

"No?"

"No." He motioned with his head over Bob Doyle's shoulder. "See that guy in there? That's Dirty Dick. *That's* a fucking fisherman." He called out, "Hey, Dick! How about giving me a hit of that vodka, there?"

A logger in Maine before coming out to Sitka in the thirties, Dick had fished in the days before quotas, regulations. On one two-day trip he pulled eighty thousand pounds of halibut in a gale and then spent the next thirty-six hours filleting it. He once drank a quart of whiskey a day, until he suffered the first of five strokes. He wore plaid shirts, jeans and Rockport slip-ons. (His second stroke made it hard for him to tie laces.) He hated taking showers. Once a week Georgia would come looking for him. *You're taking a shower now, Dick. Either that or you're out in the street.* He slept wearing his olive green cap, and only when he removed it to scratch his bare scalp would one see a few locks of gray hair, like wildflowers clinging to life on a bare rock. Like most of the hard-core fishermen, he had monstrous hands and great, hidden strength, and wore the same blue plaid shirt and jeans day in and day out. He would hide cans of Rainier in the sofa, even sleep with them, to keep the others from stealing his beer. It did no good. Every morning he would rant and rave that someone had swiped his stash.

"So," DeCapua said, "how long you staying?"

"I don't know."

"Cigarette?"

"No, thanks."

"I mean, can I get one of yours?"

"Oh," Bob Doyle said. "Sorry."

Mike DeCapua tucked a Lucky between his lips, lit it and blew a lazy, contemptuous puff of smoke.

"So where's your stuff?"

"In a couple of boxes outside."

"That's it?"

"Uh-huh."

"Hmm. Want another shot?"

"Why not?"

They killed the rest of that Popov bottle with Dirty Dick and broke the seal on another. Once DeCapua had finished his Lucky he fished a wad of tobacco out of his pocket and started rolling his own cigarettes. He also started spinning yarns. Like how he had become fluent in sign language— a necessity, since both his parents were deaf—and the first time he ran away from home at thirteen and hitched his way from Hartford to Galveston. He told them tales of his life as a train hobo, crisscrossing the country inside the boxcars of freight trains, and how he'd married three times, the last time to a ditty bitch he'd met in a motel room in Bellingham, Single-Cell Susie. Old Single-Cell had black hair that looked dipped in starch, vacant eyes, a body kept nice and trim by hepatitis C, the morals of a vacuum cleaner and an insatiable appetite for crack. In the end she bolted with a buddy of his, another fisherman, while DeCapua was sitting in a cell at Lemon Creek doing twenty-two months for breaking into the Pilothouse bar and grill down near Old Thompsen Harbor. Hadn't he heard about his Pilothouse caper? No? By God, he sure cleaned that safe—$2,346.25 in all, half of it in rolled quarters. With the money, he and Single-Cell got a room at the Cascade Inn out on Halibut Point Road. The very next night they went right back to the Pilothouse and got shitfaced on tequila, buying rounds for the house with the rolled quarters until closing. He'd even bought beers for the owner, Ron Bellows, a paraplegic from Florida who seemed like a thoroughly good guy until he prosecuted. The cops found DeCapua at the Cascade two days later, passed out in bed with Single-Cell, beer bottles and $658.75 in bills and quarters scattered all over the room. Bob Doyle just kept nodding his head and filling DeCapua's glass. The man was quite a talker, all right, but he evidently knew his fishing and seemed to have led quite a life.

The tale that Mike DeCapua told with particular relish was the story of his Idaho jailbreak and flight to Alaska in the spring of 1983. At the time he'd been making a living as a burglar—not a good living, to be honest, but at least nobody could say he was a thief without resolve. The sheriff of Latah County knew him well. Twice his deputies nabbed him in the act: on November 25, 1979, at the Log Inn, a dance hall in the town of Potlatch, and again, two years later, inside the Les Schwab Tire Commercial outlet in Moscow, the county seat. The first time DeCapua pleaded no contest to attempted burglary and got seven and a half years probation. For his second caper he received three years at the state correctional facility in Boise.

When he came up for parole, the judge asked him a series of questions that he answered with a series of promises. He promised to quit the booze. He promised to quit the smack. He promised to quit ducking his child-support. He promised to quit hitting taverns, titty bars, bowling alleys, motels and pawnshops. For good measure, he promised to read the Bible at night, to attend church on Sundays (he was technically a Catholic, though his parents had never bothered to have him baptized), to apply to seminary—in short, to become a foot soldier for Jesus. The judge, mildly impressed, not only let him out of the joint but allowed him to move across the state line to the city of Spokane. There, DeCapua got a bed at a halfway house, a job as a janitor at a Holiday Inn West and a parole officer to keep him company.

For two months DeCapua kept his PO thinking nice thoughts. Then came the afternoon he left work early and ran into some biker pals. Against his better judgment, he allowed them to talk him into splitting a pitcher of beer, and the next thing he knew he was waking up on the floor of a warehouse, lights shining in his bloodshot eyes. The bright lights, it turned out, were coming from flashlights—flashlights held by two Spokane County deputy sheriffs. The cops interrogated him for hours. There had been a rash of burglaries in the area and they thought, mistakenly, that they had their perp. But since they could elicit from him no more than a belch—

and a weak one at that—they ran his prints and photo through the computer and waited.

An hour later, in stomped the sheriff. He was a pot roast of a character, round, gristly and not at all amused looking. He slammed a folder on the table. "We know who you are," he said, and then declared rather grandly, "Mark Allen Rhodes." Actually, the sheriff was not far off; Rhodes was one of DeCapua's aliases. There was a notation to that effect on his Idaho rap sheet. But the Spokane Sheriff's Department hadn't connected the dots. Mark Allen Rhodes was, to their understanding, a local resident with an unblemished record.

Not one to pass up a chance to confound authority, DeCapua lifted his head, made puppy eyes at the sheriff and muttered, in the most ashamed tone of voice he could muster: "Okay, you got me. Gee, in a way I'm glad that you found me out. Now, sir, could I get a cigarette?"

At the arraignment the next morning the public defender stood before the judge and argued that his client, because of his clean record, should be released on his own recognizance until the trial date.

His lawyer motioned for DeCapua to stand before the judge.

"Mr. Rhodes, do you have a job?"

"Yup."

"Do you have an address?"

"Sure do."

"How long have you been living in the city of Spokane?"

"Uh, nine years."

"Your Honor," his lawyer said, "Mr. Rhodes has a residence and a full-time place of employment. And he has lived in Spokane since 1974. Considering how tight space is at the county jail, I would ask the court to consider releasing him."

The judge regarded the gaunt, young man with the long ponytail and frayed coveralls. If he had shaved, he needed another razor.

"Very well," the judge said.

Within a couple of hours, DeCapua was standing on the highway outside of town, a Greyhound ticket to Seattle in one pocket and in the other a wad of bills—mostly fives and singles—fifty bucks he'd bummed off his girlfriend. He also had a lighter, a fresh pack of Marlboros, a bag of weed and rolling papers.

By no means was this his *Great Escape* and the hounds were hardly on his heels, but Mike DeCapua was taking no chances. He was twenty-seven, and on his way to the Great Land.

Life on America's last frontier, he quickly learned, wasn't all greatness. The rest of that spring Mike DeCapua spent scraping cod guts out of the holds of longliners. When he got sick of that, he moved to Juneau and got a job installing drain tile in bathrooms. When he got sick of that, he found work flame-broiling steaks at the Black Angus.

It rained month in and month out. There were few taverns and even fewer tire dealerships to hit in a pinch. Pot was expensive. Beer was no buy—cheap brew went for four dollars a mug, six in the strip clubs. And there weren't many tits in Alaska beyond those on display in the titty bars.

He also had to part with a number of cherished myths about frontier living. His second summer in Juneau he made his first solo foray into the wilderness, a two-week trip. Into a twelve-foot canoe he packed a fishing pole, a crab trap, a .4470 Winchester live action, a .45 automatic, oatmeal, flour, coffee, tobacco, beans, salt and baking powder—for his smelly feet. Finally he bought something he'd always wanted: a tepee. It took him nine hours to put up, nearly five to take down. After that, Mike DeCapua went back to pup tents. He had to give it to white America on that one. Pup tents had a practical upside.

Of course, there were other upsides. He liked watching eagles dive on old bait the hand trollers dumped in the channel after a trip. He liked catching a halibut and selling it for a hundred and forty dollars. He liked the fact that during his first three months in Alaska no one had asked him to sign anything. He liked that people could tote a gun without drawing a second glance. And he especially liked that everyone he met hated the government.

Surrounded by the raw material of life, Mike DeCapua nonetheless found it difficult to remake himself. Like an athlete breaking training, he resumed partying and went about screwing as many Native women as he could impress with his tales as a mountain man. During one interval he did date a redhead, Raye, and even moved in with her and her daughter. But one morning after a minor spat he simply packed a duffel bag and walked to the police headquarters in downtown Juneau. He told the officers he wanted to turn himself in for transgressions in Idaho. When they did not take him seriously, he told them to call the Latah County courthouse. They did, and were connected to a judge named Schwam.

The judge was succinct.

"If he doesn't like the snow in Alaska and feels like turning himself in, tell him to come down here and do it." The judge paused. "He got up there on his own dime. He's not coming back on mine."

So he stayed. First he moved in with four other Lower Forty-eight castaways who lived on Eaton Street in a trailer, which they called the Den of Doom. It was a den of drunks, actually, a mecca for displaced souls without a bed or a boat for the night. Their beer, acid and coke parties raised eyebrows, especially those of the pastor of the local Assemblies of God church, which shared their driveway. On Sunday mornings after the church doors opened, an entourage of loggers, miners and fishermen descended on the place, each with a bottle of cheap whiskey, gin or vodka under one arm and a case of Miller or Bud on their heads for another NFL Sunday.

After football season ended, a foreigner came to stay a few nights. He was a little man, kind of dark, kind of hairy, with clothes that always reeked of pipe tobacco. Birol, he called himself. The man was from somewhere in Turkey, and wanted very much to stay in the United States. He was even willing to pay for an American wife.

"Buddy," DeCapua told him, "for ten grand I'll get you the woman of your dreams."

The dream lady was an old girlfriend of his from Spokane, the one he'd left behind when he bolted to Alaska. Robin Germen was her name. He remembered her as being either nineteen or twenty, with a neat, waifish little body, disheveled wisps of floating brown hair and a head full of fluffy dreams. One of those dreams was to go to Alaska. Then he remembered she had two kids.

Well, he thought. Nobody's perfect.

He hadn't called the girl in two years and had lost her number, but remembered her mother lived in Eugene, Oregon. He left his phone number, and not long afterward, Robin called him back. It took some doing, but he finally convinced her that he really did miss her, after which he sent her and the kids one-way plane tickets to Juneau.

He greeted them at the airport, put them in a cab and took them straight back to the Den of Doom. She had barely put down her luggage when DeCapua asked, "Say, would you marry a Turk?"

She laughed at that. So he repeated the question. She stopped laughing and asked if he was joking. He said not really. She told him to cut it out. He said the Turk was a nice guy. She told him to go to hell. He said the money was nice. She told him to really go to hell and stay there for a while. He said take it easy, put that suitcase down, calm down, calm down. She didn't say anything for a while. He said pardon me and then grumbled, hell, I was just thinking we'd make a few bucks. That's all.

Since he had lured Robin up to Alaska, Mike DeCapua felt it only fair that he marry her, which he did shortly thereafter. Then he bought a boat. It was a dory, twenty-seven feet long, flat-bottomed, with flared sides, and in need of some sanding and paint. The trailer was no place for kids, he reasoned, and so one morning he piled his family into the *St. Pia* and set sail for Pelican, a clump of cabins at the toes of a fjord, a skip from Glacier Bay.

The country was as grand as it had been in his dreams. On both sides of the Lisianski Strait the mountains stood up proudly, clothed in dark, jagged spruce whose fragrance came out in the drenching rains. The trees changed their character many times during the course of a day; at times they seemed to brood, at times they waved and shone like running water, or a candle's flame. Sometimes they felt quite close and other times, quite far away. There were trees that looked like the double sails of an old schooner and others like the serrated forms of black cod. On clear, cold nights, when the light was dying, it was as though a thin, gold line had been drawn along the tops of the mountains. Finally, when night fell, the fjords looked like immovable black waves against the sky, and every sound— every murmur of the channel, every beat of a bird's wings—seemed to travel a great distance.

If the art of moving gently, without suddenness, is the first to be learned by the hunter, then Mike DeCapua learned the benefits of patience by watching humpbacks seine herring in the channel. The silvery giants would glide up on their prey in a wide circle and time their burst from the water, seemingly taking the channel up with them as they rose, true silver, flukes unfolding and spreading in flight, the spray on them sparkling as though they were studded in diamonds; and then falling back with a splitting clap, the whales would throw geysers of water that stunned the herring and made them a cinch to catch.

And so it was there, in Pelican, that he found his one, true calling—fishing. After a trip in which he caught nothing and lost all of his gear, Mike DeCapua took a job as a deckhand on a longliner owned by a Swede named Ing. Together with a man who called himself Barry the Englishman, they fished the nooks and crannies up and down Alaska's southeast archipelago and, later, the open gulf, in search of gray cod. That they regularly went three days with no more than an hour's shut-eye each day meant little to him; for the first time in his thirty years, he loved his work. His thoughts did not jump around. His breathing came easier. He did not feel cornered.

With time, Mike DeCapua grew into a solid deckhand. Coiling was his forte. He could sit at a shiv, wrapping up incoming longline for thirty-one hours without a break, and after a two-hour nap do it again. It was tricky business. Many a coiler had lost eyes, noses, fingers and ears to leaders and hooks gone awry, and he had his share of close calls, too. But coiling seemed to fit DeCapua. He was fast. He was so smooth that Ing kept him on for three years. It improved his reputation as a deckhand, though it did not do his marriage any good. After three months of no Mike, Robin packed her bags and put herself and the kids on the ferry to Sitka. Winter was coming. She was pregnant with his son and she'd had enough of sleeping on a dory. He returned from one of his fishing trips to find the *St. Pia* empty, bumping against the dock.

One day a letter arrived in his post office box. He read it over carefully and showed it to Ing. Idaho wanted him back. It had taken the authorities three years to extradite him but they had finally gotten around to doing it. Apparently, he'd been sentenced in absentia to eighteen years for jumping parole. The Alaska authorities were to return him at once for a hearing.

A month later, Mike DeCapua was back in Boise. Before his hearing began at the Idaho superior courthouse the judge called him into his chambers.

"They tell me you're a fisherman," the judge said.

"Yes, sir."

"Like some coffee?"

"Sure."

The judge poured and then sat back in his leather chair and considered him.

"You know," the judge began, "I'm an angler myself."

"Sir?"

"I've always wanted to fish in Alaska. Tell me, now, from one fisherman to another—is it as good as everyone says?"

If Mike DeCapua was good at anything, it was telling people what they wanted to hear. And he really heaped it on this time. At length, the judge sat forward, his bristly eyebrows bunching.

"Tell me something, Mr. DeCapua," the judge said. "What will you do if I let you out?"

"I'll go fishing in Alaska, Your Honor."

"I have your word on that?"

The sentence was commuted to a year, and after ninety days Mike DeCapua was released for good behavior. The police escorted him to the airport. They wanted to make sure he was on the plane when the cabin door shut.

His ticket to Sitka was one-way.

S E V E N

The couch at Georgia Kite's remained Bob Doyle's refuge through the rest of November. The days grew shorter and the nights colder still. The snowlines on the mountains were lower with each dawn. Before daybreak a frost would leave thin ice on the porch steps and whiten the branches of the big tree out back. There were no flowers to see anymore, but by late morning the frosts would go as if they had never come, leaving beads of liquid like tears welled up on the leaves of bushes and on the fallen spruce in the forests.

Then the solstice came and went, and Christmas, too. It had been snowing in the mountains. Rangers closed the access road to Cascade Park a third of the way up the mountain because of the drifts, but hikers were still allowed to roam about the Old Mill Site out toward Herring Cove. After a heavy rain the trails would be softer to walk on. But if the air hardened with a cold front, the pleasure of feeling the forest give a little under his boots would not be there and it felt as though he was walking down a damp gravel driveway or a field of frozen crickets.

By now Bob Doyle had met most of the flophouse mainstays, though there was always another newcomer who had fallen among them. Harvey Kitka and Spoon Davis still had enough of their summer fishing shares left over to keep the beer kegs filled. Dirty Dick was still fighting his weekly showers, though he'd given up trying to hide his beer, and Sue Nelson, who lived four doors down, was still fighting with her husband, Perry, about her drinking. She was in the habit of stopping by the Basement at night, wearing tight sweaters and laughing that smug little laugh of hers and making eyes at John Gino so he would let her swig his vodka. A homosexual had made a pass at Dirty Dick, which in turn threw Sue Nelson into an uproar. A couple of sisters from Jackson College had been trying out for size most of the fishermen in the Basement bedroom. Rob Kite was still bringing weed up from down south, but having a hard time undercutting his competitors.

It was a slow time of the year for fishing, and Mike DeCapua was getting a little cranky about having to stay put in port and do boat repairs—busywork, he called it—while his skipper waited for a window of clear weather between blows. The boat he'd been on, a longliner called the *Min E,* hadn't been pulling much and he was in an increasingly evil mood. He was in a very evil mood the afternoon he banged down the stairs of the Basement and found Bob Doyle shooting a rack of pool with Norm Niessen, a friend who lived in the woods out off Sawmill Point Road. Niessen had moved up to Alaska from South Dakota years earlier. He was describing what it was like to live five years in a dilapidated school bus.

"Is he still talking about his days in that fucking bus?" DeCapua said.

"Kiss my ass, Mike," Niessen said.

"Sorry, Mom." DeCapua turned to Bob Doyle. "It's just that I've heard that one about a zillion times."

"I said kiss my ass," Niessen said.

DeCapua smiled. "You know what I call this guy, Bob? I call him Perfect Mom. Tell him, Norm. Tell Bob here why I call you that."

"Quiet, Mike," Bob Doyle said. He was leaning far over the table, lining up his next shot.

"Listen," DeCapua said. "All you dumb fucks ever do is shoot pool. Fuck pool. Say, Bob, do you want to play pool all your life?"

"You know something better?"

"How about fishing?"

DeCapua told him that one of the deckhands on the *Min E* had gotten into a scrap with the skipper over holiday time. The skipper was not letting crewmen go for the holidays. So the deckhand split.

Bob Doyle looked at DeCapua good and hard.

"You're shitting me."

"I ain't."

Bob Doyle lowered the cue stick.

"When does the boat leave?"

"New Year's."

For two days they untangled snarls in longline gear, stocked up on fuel, groceries, and tweaked the engine. Before daybreak on the first day of January 1998, the day of a rockfish opening, the *Min E* sailed out of Sitka Sound.

It sailed right back one night later. The rockfishing was horrible; after thirty-two hours of haulbacks, they'd barely pulled three hundred pounds.

Bob Doyle was not nearly as upset as the others. He had made it through his first commercial fishing trip without a major screwup. And he felt he had hit it off with the skipper, Phil Wiley. After they had tied up, Wiley walked up to him and invited him to stay on another month, and to sleep on the boat, too, if he liked.

"Phil's got a crush on you," DeCapua said. He and Bob Doyle were walking back from the ANB to the Basement. "What you do, anyway? Give him something special in his rack?"

"He's a good guy."

"He's a cocksucker."

Wiley, he explained, had shortchanged him out of three hundred dollars after a trip the previous year. Once they had beached and the rest of the crew had split their shares and gone to the Pioneer Bar, DeCapua pulled Wiley aside.

He told him, "Okay, I don't care what you do to the rest of the crew. They're not here. It's just you and me."

"Go on."

"You owe me three hundred dollars. I want my three hundred dollars. You give me my three hundred dollars and I keep my mouth shut and go home. You don't give me my three hundred dollars and I'm going to the crew. First I'll tell them what a retro check is. Then I'll tell them you owe them a retro check. Don't think I won't. I got all the catch figures. I want my money and I want it now."

Wiley looked blankly at him. "Is that it?"

"That's it."

"Fuck off."

So Mike DeCapua marched back to the Basement, pulled out a phone book and called every marine insurance company listed in the yellow pages until he found Wiley's. He told the agent that his client was a high risk.

The agent asked him why.

"Well, your guy is not paying the crew. Which makes the crew mad. And they're going to retaliate by hurting the vessel when Mr. Wiley is not around. And you are the ones who are insuring him."

"I see."

"You guys are going to wind up paying that boat off to the bank because it's going to get lost at the dock sometime because of his behavior."

"I see," the agent said.

"And what's more, that vessel ain't safe. Her mast is unsteady, her deck is loose from the rails, and she needs extensive hull work. And he ain't doing it."

Three days later, a marine architect showed up to inspect the *Min E.* He took some notes, made out a report and sent a letter to Phil Wiley not long after. It said that unless repairs were affected, the boat would no longer be insured.

The repairs, DeCapua said, set Phil Wiley back a bundle.* Still, he did not fire DeCapua.

"But he sure made me pay," DeCapua muttered.

"How do you mean?"

"Well, when Phil takes me out, we work just enough to get a small catch, to cover some expenses he's got, and then we come right back in. There's never enough left over for me to get a good paycheck."

"Oh."

"He can be a bastard, all right."

The *Min E* sailed again on January 15. Bob Doyle and Mike DeCapua were on it. They worked the lower half of the Chatham Strait and returned to Sitka with one thousand pounds of black cod.

"That bastard," DeCapua muttered. He and Bob Doyle were walking up the dock. "That fucking bastard did it to me again."

"Listen, Mike," Bob Doyle said. "I don't know if he's doing it on purpose. I mean, I don't see the sense in it."

"In what?"

"In shooting himself in the foot," Bob Doyle said. "I mean, why would he do that?"

Mike DeCapua laughed.

"People do it all the time," he said.

---

*In a phone interview in November 2003, Phil Wiley said he vaguely recalled a dispute with Mike DeCapua over retro pay and said his deckhand did contact the company that insured the *Min E* to lodge a complaint about the boat. However, Wiley said any allegations that his vessel was not seaworthy or that he slighted crewmen of pay were false. The skipper said the *Min E* was a fine boat with a sturdy hull that required only routine maintenance, and that at no time could he recall any insurance inspectors finding otherwise. Though he could not remember any other details about his disagreement with DeCapua, Wiley did say it was not uncommon for skippers in Alaska to hold back the retro pay of deckhands who jumped ship or left a vessel before it had been adequately cleaned and retooled after a fishing trip.

The morning after the *Min E* returned to Sitka from its second bad trip in as many weeks, Bob Doyle went to the Halibut Hole to look for Mike DeCapua. He was not there, so Bob Doyle sat down and had a coffee and a black bean soup with cheddar crackers. He read the previous day's paper with the coffee and smoked a cigarette. It was raining out and there were large, oily puddles in the street. Two young women came out of a hairdresser's shop next door, touching their necks and laughing, and picked their way around the puddles. More people went by the window on the sidewalk. Most wore slickers and boots. A man was carrying a little girl on his shoulders. The way the girl's blond bangs stuck to her forehead and the terribly bright smile on her face sent a chill through Bob Doyle so he ordered another cup of soup with crackers and lit another cigarette.

Mike DeCapua did not turn up, so around eleven-thirty Bob Doyle walked up the main street to the Moose Lodge and had a Crown Royal with the barman. DeCapua didn't show up there either, so he left a five dollar bill on the bar and thumbed a lift out along Halibut Point Road to a yard where shipwrights welded the hulls of steel boats. He knew that DeCapua used to camp out at the yard, sleeping inside the hulls of ships that had come in for repairs. He asked a couple of the metalworkers if they had seen him. They shook their heads. Nobody had seen DeCapua for months.

In the rain Bob Doyle walked back along Katlian Street. The plywood of the sagging houses looked gray and wet. He turned in at the ANB and was about to take the ramp down to the docks when he spotted Mike DeCapua coming across the parking lot. There was no mistaking his heronlike strut. Bob Doyle called out to him. DeCapua came over and stood next to him under the awning.

"What's wrong?"

"Nothing," Bob Doyle said. "Where you been?"

"At the bank."

Bob Doyle gave him a puzzled look.

"Restitution," DeCapua said. "I pay fifty bucks a month to this joint I hit a while back, the PO leaves me alone."

"Oh," Bob Doyle said. "You won't believe what I did today."

"What?"

"I think I may have found us a job."

"No kidding."

"The job service called me this morning. They got a job for us."

"Baking cookies?"

"No," Bob Doyle said. "Somebody posted a job today for two deck-hands on a fishing boat."

"Two hands?"

"I got it right here."

He felt in his pocket, pulled out a piece of paper, unfolded it and handed it to DeCapua.

> OPEN TIL FILLED. SEASONAL DECKHAND POSITIONS FOR ROCKFISH SEASON. QUALIFICATIONS: PREFER PRIOR DECKHAND EXP OR SOMEONE WHO HAS SOME KNOWL-EDGE OF BOATS. LONGLINE STUCK GEAR USED. WILL BE FISHING THE SITKA SOUND WATERS. WORK MAY EXTEND TO TENDERING FOR THE TANNER CRAB SEASON. WORK IS TO BEGIN IMMEDIATELY.

"You got a number?" DeCapua asked him.

"I already called."

"And?"

"The skipper said he'd need two guys. He said he'd take one green hand if another experienced one came along."

"He did, huh?"

"This could be what we're looking for. And the best part is that we'd be set for February, too—just tendering fish from the other boats."

DeCapua was reading the ad again.

"It doesn't give the skipper's name," he said.

"It's Morley. Mark Morley."

"Never heard of him."

"What does that mean?"

"Nothing," DeCapua said. "Means I ain't never heard of him, is all."

"He's down at Old Thompsen Harbor. He said he'd be there pretty much all afternoon. If we went down there now we'd probably catch him."

DeCapua was reading the ad again.

"And what's our take?"

"Ten percent of the catch."

"Ten each?"

"Yeah."

"That ain't bad," DeCapua said. "But what the hell kind of skipper gets his deckhands from a newspaper ad?"

"It wasn't in the newspaper," Bob Doyle said. "I got it from the job placement service."

"Same fucking difference."

"Well," Bob Doyle said, "it could be just what we need to get us through the winter."

A smirk flickered across DeCapua's lips. "It would be something to see Phil's face after we tell him that we're splitting."

"So we go?"

Mike DeCapua looked out at the channel. There was a wind rising, a chop building.

"Oh shit," he said. "What do we got to lose?"

It was getting on evening when they reached Old Thompsen Harbor. The wind was ripping along the breaker wall and the tide was up. They walked down a grated ramp and under the orange glow of the lamps along the main pier. There were not many boats in their slips, but the few left there for the winter were rocking and pitching in the chop. To the left, down the far end of the third dock, they saw the silhouette of a long, dark ship and the dark shapes of two men on the foredeck.

"That must be it," Bob Doyle said.

"It's a schooner."

"Is that bad?"

"No," Mike DeCapua said, "I like schooners."

They walked along the dock until they came to a ship with a rounded, black stern, with the words LA CONTE painted on it in white. A heavyset man in a slicker and a cap was standing on the slip, handing a bag up to a second man on the bow.

Bob Doyle called out, "Howdy!"

The big man on the dock turned.

"You Mark Morley?"

"Yeah?"

"My name is Bob Doyle. I called you this morning. About those deck-hand jobs?"

"Oh, right!"

The man came over, looking awfully big in a sweatshirt two sizes too large for him, and a cap and slicker. He was not tall as much as he was stocky and bull-shouldered, the size and build of a tailback gone a little heavy. He wore his cap cocked way up with the brim curled, like a kid in Little League might, and he had big, floppy ears and thin, wide-set eyes that seemed small behind his thick, oval glasses. Bob Doyle thought he looked like a philosopher in a fisherman's jacket.

"This here is Mike DeCapua."

"Good to meet you," Morley said.

"Same."

"So, you guys looking for work?" Morley asked.

"Depends on what the work is," DeCapua said.

"We do gray cod, mainly," said Morley. "We'll probably take some dog sharks, too, being that there's so many damned sharks and that nobody ever misses a shark." He laughed.

"Who's he?" Bob Doyle asked, motioning to the man on the foredeck.

"That little guy? Oh, that's just my manager. What I call him, anyways. His name's Gig. Hell of a fisherman. You know him?"

"No," Bob Doyle said.

"I know him," DeCapua said.

"Hey, Giggy," Morley shouted. "Come on over and say hello."

The man walked to the stern of the boat and leaned over the bulwark. He was Native; that Bob Doyle could see straightaway. He was small in height, but rangy and strong looking, with thick, stubby fingers. His head was covered with the hood of a sweat jacket, but out of the sides of the hood stuck tufts of hair, shiny and black as a raven's wing. But for his mustache, there was something boyish, puppylike, about his face. It might have been the smooth, fair skin, it might have been his eyes, black and long-lashed and wetly shining, like a harbor seal's. He also had this crooked, shy grin, like a schoolkid who's just been busted for smoking in the bathroom.

He was wearing one of those grins.

Morley continued, "This here's Bill Mork, but everyone calls him Gig." Morley pointed to Bob Doyle. "Giggy, this here's Bob, and that there's Mike Dee . . ."

"DeCapua."

"Thanks."

Mork said, "I know him."

"You do?" Morley said. "Well, isn't that nice. Just like family." He laughed. Then, to Bob Doyle, he said, "Giggy knows fishing. Been at it his whole life. He's from Pelican. He knows his way around the waters here."

Bob Doyle smiled.

Morley turned to DeCapua. "I hear you're a pretty experienced fisherman, too."

"You heard right."

"Who you been out with?"

DeCapua mentioned the names of skippers and vessels he had fished with, but Bob Doyle could see Morley did not recognize them.

"How'd you get started?"

"On my own boat."

"Your own boat?"

"Then I realized I was a better deckhand than a skipper, so I went to that."

"What're you good at?"

"I can coil. Bait."

"You know your way around a boat?"

"I know how to make a set. I know what everybody's job on deck is."

Morley pushed up his glasses. "I'll bet you do. I'll bet you do." He looked up. "Giggy, you say you know this guy here?"

DeCapua and Mork traded nods. They were not friendly nods, but not cold ones either.

"Giggy and me go back some," DeCapua said. "We fished all the way from Ketchikan to St. Paul. And man, that fucking St. Paul is a great place to party. You ever been there?"

"No," Morley said.

"Well, shit, you ought to go. There's a Native woman behind every tree on that island. Only one problem with St. Paul—it ain't got no trees."

They all laughed.

"So what kind of fish you like to catch?" Morley asked DeCapua.

"Depends on where I go fishing," DeCapua said.

"Well, we've been doing mostly inside waters," Morley said. "Chatham. Peril Straits. Tenakee Inlet. Freshwater Bay. Whitewater Bay. Shayek. Places where gray cod hang out."

"Right."

"I bet you've done some rockfishing, too," Morley said. "I'd be willing to put money on that."

"Done my share."

"I'll bet you have," Morley said. He paused for a few seconds, and then he said to DeCapua: "All right. I'd be willing to take you and your buddy, Bob, along with us and give you both a ten percent share of our catch. We'd be leaving day after tomorrow. Probably be gone a week. Of course, if you guys do good work, then I might keep you on through February to do some tendering."

"Sounds good."

"Great," Morley said. "There's just one thing, then. Old Bob here is an old Coastie. Now that's all right, but he's green on a fishing boat. He's going to need some watching. You'd be willing to keep an eye on him, am I right?"

"Sure."

"Good. That's good. All right. Anything else you want to know?"

"I'd like to see the boat."

"Sure," Morley said. "I'd be happy to introduce you to the old lady."

As schooners went, she was no beauty. The hull was bruised and loose in spots. The decks were worn, buckling under the weight of too many booms and too much rigging. Worms were eating the frame around the rudder. The hold hatches leaked. Mildew grew up and down the cabin walls and made black spots on the ceilings where water got in. The head was stopped up. The lazarette was a shambles. Water had to be pumped from the bilges every couple of hours, even in dock. The engine room looked as though it had been taken apart and never quite put back together.

Mike DeCapua pulled himself up out of the front hold, clapped his hands and rubbed them on his sweatpants.

"Well?" Bob Doyle asked him.

"I like her," DeCapua said. He pulled out a pinch of tobacco and rolled himself a cigarette.

"Why?"

"She sits firm in the water. She ain't tippy."

"Oh."

"She was set up as a buyer boat," DeCapua said. His eyes skipped around the foredeck. "She's tired, all right. I hate to see a boat get treated bad. And this girl's been treated bad for a long time. But she's solid. Notice when I go walking over here, or walking over here? See how she just stays where she is?"

"Yeah?"

"That's a good boat."

"She needs work."

"All boats need work."

"I guess."

"She's got soul," Mike DeCapua said. "She don't tip and she's got soul." He pinched his cigarette, lit it and looked vaguely up at the rigging. "Fucking people don't know how to treat a boat, is all."

She hadn't always been so tired looking. The day she came out of the Tacoma shipyard, she had the simplicity and sleekness of a kayak and the sweet scent of Washington fir. The G.W. Hume Company, her first owner, named her *Narada*. That was in January of 1919.

Though quite slender, she was seventy-seven feet long and built as sturdy as a pocket battleship. Her deck planks were two and a half inches thick. Her beams were seventeen inches in width; her frames and cross-beams, eight inches square. Her hull, two inches of creosoted, vertical-grain fir, had been overlaid with an inch of ironwood to keep her from bruising on docks. At sixty-six tons—with her holds and fuel tanks empty—the *Narada* was no lightweight, and sat low in the water. Whenever she sailed full of fish or tanked down with ice, she looked like a surfacing submarine, her decks constantly awash in frothing brine.

Through inside waters she moved like a saber, her high, up-curving bow handily slicing the chop. But her slender figure betrayed her on the open seas; on the great swells of the gulf or the Pacific she would heave and wallow, and when big combers broke against her hull the full length of her would shudder—a shudder that brought a pallor to the cheek of the staunchest deckhand.

Originally, she had been commissioned to tender fish caught by trap vessels to packinghouses in Ketchikan and Petersburg. In the twenties Alaska fishermen used nets framed in fir and cedar, anchoring these traps at the mouths of bays or leaving them to float in channels or near rivers. They worked so well they wiped out salmon and cod stocks up and down the Inside Passage. By the early thirties the traps were outlawed, leaving most trap tenders, including the *Narada*, sitting in dry dock, as jobless as the rest of the country.

From then on, the boat was revamped, remodeled and refitted so many times for so many different jobs and owners that she lost some-thing—her identity, some put it. During the gold rush of the thirties, the Dupont Dynamite Company chartered the *Narada* to haul barges of ex-plosives to Skagway. During World War II she tugged weapons and military cargo. In the seventies she was converted into a tender, only to be recon-verted a decade later into a longliner. Her two original holds were replaced with diesel tanks, which were in turn changed to fiberglass-lined tanks, which were later ripped out in favor of refrigerated holds. Dupont upgraded her engine to a 425-horsepower, Cat-343, but the mechanics dispensed with the muffler to make her go twice as fast as the speed she was built to go, seven knots. A third fish hold was put in, along with a twenty-ton compressor and a new network of wiring and three-inch piping to carry chilled seawater to the holds. Different owners updated her navigational instruments at different times. Her original, spoked wheel was swapped out for a stainless-steel model. Three new cargo booms were added, followed by a second mast, an extra set of rigging, new bulwarks and a bait shed. (Although it could now load fish simultaneously from four different vessels, the extra weight on top made it harder for the boat to recover from rolls.)

Not even her original name lasted; in 1971, she was reregistered as the *La Conte*—the Count.

In 1987, a tugboat broadsided her at the mouth of Wrangell Harbor, and the *La Conte* sank faster than a packed crab pot. She was raised, beached and put up on the gridiron. It didn't look good. The schooner had lost thirteen hull planks, seven frames, eight beams, six deck planks, and there was extensive damage to the structure that fastened the galley to the deck.

After eighteen months of litigation, the boat went on the market and sold, as was, for twenty thousand dollars. The buyers were the only people who made an offer, Jeff and Shannon Berg, a young Petersburg couple with fishing experience and romantic notions about aging vessels. They restored the hull and deck with the same, aged Washington fir the boat's builders had used, gave the *La Conte* a top-to-bottom painting and put her to work. For ten years they tendered together up and down southeastern Alaska, as far north as Prince Rupert Sound, sometimes even fishing halibut, black cod and salmon.

As it turned out, the boat lasted longer than their marriage. In the spring of 1997, as they waded through a divorce, the Bergs put the boat up for sale. Within a month, a buyer came along.

Unlike the Bergs, the *La Conte*'s new owner was no fisherman, and no more Alaskan than a keg of nails. His name was Scott Echols, a plump, smooth-tongued salesman from Empire, Georgia, who had made a small fortune in suckling pigs and goats.

Echols specialized in slaughtering his goats halal—in such a way as to be ritually fit, according to Muslim law. It was a niche market and Echols had it locked. At the top of his game his company slaughtered three hundred goats a day. To a friend he once joked that he must have killed every goat within a hundred-mile radius of the Georgia line. Echols never did figure out why Muslims made such a fuss about getting their goats halal. But it mattered little to him. He was no Muslim. He was a businessman. The goat king of Georgia. And he had the bank account to prove it.

How he made the jump from halal goats to Alaska seafood was, like all of his ventures, something of an accident. Out of college he went to Hawaii to get a master's in robotics, then was hired by Arco to be an analyst in Anchorage. He quit his job and started his own robot company, WARP Industries. The company received a lot of state grants, was written up in a lot of magazines and in two years barely broke even. Alaska, he had learned the hard way, wasn't ready for robots.

So Echols took a job as a systems analyst for a Japanese fishing conglomerate. Day after day the company's accounting figures passed across his computer screen, and each day the numbers left him speechless: his employer had been buying Coho salmon at $15 a pound and selling it for $250 a pound. He was so excited by the balance sheets he quit his job and tried his hand at fishing. That was not a good idea. He got seasick and couldn't give away the bait at the end of his line. In 1992 he took his master's degree in digital communications engineering and flew back to Empire.

After three years of dealing in pigs and goats, Echols returned to Alaska in the fall of 1996 to start his own company, World Seafood Producers. He had since married his childhood sweetheart, Cherie, whom he had met when she was four and he eleven, and had picked up front money from some Japanese and Korean investors he'd met during his robotics days. His latest idea: to sell to Japanese and South Korean buyers the part of salmon everyone else in Alaska tossed in the garbage — salmon roe.

He set himself up in Juneau in a lakeside duplex and shopped around for a cheap tender to move his product to cold storages in Bellingham. He drew his share of looks. At the University of Georgia he had been a second-string cornerback, but now Echols was somewhat beyond athletic weight, with the shoulders and dignified gait of a grizzly. He had a small, round head, eyes the color of blueberries, a pug nose and a quick grin. He wore

European colognes, crewneck sweatshirts under ski jackets, Levi's 505s, designer hiking boots and a blue beanie.

The first thing he heard about the *La Conte* came from a cod fisherman in Juneau, Fred Damer. Damer had a friend, Jeff Berg, who wanted to unload an old tender to pay for a divorce. "You won't find another boat that big for that cheap," Damer told Echols. In the end, a final price of $109,000 was agreed upon. Echols turned up at the closing $4,000 short, however, so Berg kept the boat's sideband radio and Uniroyal life raft—items Echols never bothered to replace.

Echols put Damer in charge of making the *La Conte* seaworthy; a month later he fired him. Damer sued for severance pay. When Echols didn't show up for the proceeding, a judge ordered him to compensate his former employee to the tune of $8,578.77, mustering-out pay, as it were.

And he soon realized there were more bills coming. A Petersburg shipwright told Echols it would cost eighty thousand dollars to fully repair all of the vessel's hatches, cracked frames, decking and hull. Echols nodded and told him to lead-patch the loose, worm-eaten planks on the stern. Then he told him to caulk the grid and add an aluminum bait shed on the aft deck. There went thirty thousand dollars.

Everything else would have to wait.

By that time it was nearly summer and Scott Echols wanted his boat to start making money. He turned to a man he'd met the previous winter in Seattle, Rob Carrs, to skipper the first boat of his dream fleet.

Carrs was college-educated, a New York native with big-city savvy who had moved to Seattle in the eighties to live the Alaska adventure—part-time, summer adventures. Carrs was good. He knew boats, he had sea smarts, and he had the right palaver with fishermen and Alaska natives. He had never seen the *La Conte*, but told Echols he'd accept the job for $30 an hour. Echols offered $26.25—all in cash, all off the books. Carrs muttered and took it.

Later, he almost wished he hadn't. Figuring out how the boat worked and overhauling the engine took three months. The *La Conte* did not go out again until September, when Carrs took her tendering between Sitka and Juneau. There were no worries—until the day in October that Carrs filled all three of her holds with chum salmon.

Halfway across the Chatham Strait, Carrs noticed the sluggishness: the engine room was filling up fast with water. Hurriedly, he connected a hose to the powerful Maxi-Flow that circulated refrigerated seawater through the holds and began pumping water in a six-inch-wide stream over the side. It took him more than an hour to get things under control.

In port, he traced the leak to several loose planks around the fantail.

A week later, after one final longlining trip near Petersburg, Carrs returned the *La Conte* to Sitka and told Echols he was quitting. The boat was an icebox, he said, he was sick of repairing everything, and there wasn't a single dry bunk on board. The two men settled money matters, shook hands and wished each other luck.

Nothing was said about the water problem Carrs had eight days earlier.

In mid-November, Scott Echols got a call from Mark Morley. Rob Carrs had introduced them eleven months earlier in Juneau, before taking Morley along as a deckhand on a black cod trip. Morley told him straight off that he wanted to take the *La Conte* out rockfishing.

"Rockfish?"

"It's the only fishing this time of year," Morley told him.

Echols hesitated.

"Listen," Morley said. "There's a bunch of two-day openings in December all around Baranof Island and a big opening on New Year's Day."

Echols was listening now.

"Yellow eye is getting a good price," Morley continued. "It's close to two bucks a pound now. That's better than black cod."

"Is that right?"

"Check around."

Echols told him he'd think it over, and he did. He remembered that Mark Morley had tidied up the pilothouse, rewiring the electronics, installing a new dashboard, stripping layers of ancient paint off the cedar woodwork. And he hadn't demanded a penny.

That night Echols called Rob Carrs at his Seattle apartment.

"What do you think about Mark?" Echols asked. "He wants to take the boat out rockfishing. Give it to me straight. You think he's up to it?"

"No," Carrs said.

"No?"

"No," Carrs said. "I wouldn't hire him. Not for this boat."

"But he's your buddy."

"You wanted it straight," Carrs said. He paused. "I like Mark. He's not a mandy-pandy guy. He's gung ho. He wants to be a skipper. But I don't think he could park the thing, let alone drive it."

The following night Echols and Morley sailed out to the Icy Strait. The chop kicked up. Echols told Morley to cut the motors and to pull in the skiff, which they were towing.

"Relax," Morley told him. "It'll be all right."

Not five minutes later the retainer broke. The boat skipped away in the darkness. A day later, Echols saw the skiff sitting in his neighbor's driveway in Juneau. He phoned Carrs again. Raving.

"Whoa, whoa, whoa," Carrs interrupted, "I don't need to hear this."

Echols sighed. "Look. I need to do something with this boat. I need it to start making some money, only I don't have many options."

"Well," Carrs said, "if you don't have many options, I guess you gotta hire him, right?"

"I guess so," said Echols.

T E N

They were standing on the dock, looking at the hull of the boat shadowed against the dark. Gig Mork had already gone back to stowing ice in the holds.

"So," Mike DeCapua said, "when are you planning on heading out?"

"Midnight tomorrow," Mark Morley said. "But I'll need help getting her ready before then."

"What's there to do?"

"Got lines to check, hooks, fuel, maybe thaw some bait. And then there's the motor, too."

"That's a full day."

"Yeah."

"And where were you thinking of fishing?"

Morley told him he planned to drive north and east up Peril Strait to Chatham, then south along the back side of Baranof Island and straight on down to Coronation Island. He wanted to fish the shoals west of Coronation.

"Then what?" DeCapua asked him.

"That's a lot you want to know," Morley said.

"Well," DeCapua said, "I like to know who my dance partners are."

"I bet you do," Morley said. He dug a finger in his ear. "All right. If Coronation doesn't pan out I figure to try the shelf along the Hazy Islands."

"The outer shelf?"

"Uh-huh."

"Say, Mark," Bob Doyle said. "You don't mind me asking you something?"

"Go ahead."

"You been skippering on this boat since November, isn't that right?"

"That's right."

"What happened to the rest of the crew?"

There was a hung instant of silence, heavy as thunder.

Morley said tightly: "We had one other hand. But he was a son of a bitch." The white cornea showed all around the tobacco-colored iris of his eyes. "A real son of a bitch."

"How so?"

Morley gave Bob Doyle a dark glare.

"The guy sabotaged the wires to the bilge alarm because it was going off all the time. Drove him crazy, he said. Fuck him. You don't go cutting wires like that. Not without talking to me first."

"No," Bob Doyle said.

"So I told him to walk."

"I see."

"So," Morley said, "are you guys in or not?"

"Can we let you know in the morning?" DeCapua asked.

"What time?"

"Nine o'clock. We're on the clock at eight and we already got a skipper, you know."

"Well," Morley said, "call me in the morning, then. I need to know either way."

"That's a ten percent share for each of us?"

"Ten percent."

They went up the pier and under the awning along the entrance ramp and out across the gravel parking lot. They looked back. The schooner looked small now in repose.

"You know," DeCapua said, "that guy ain't done a whole lot of cod fishing."

"How do you know?"

"Let's just say I've fished with guys who ain't done a whole lot of cod fishing."

They kept walking.

"I guess we'll stick with Phil, then," Bob Doyle said.

"Like hell we will," said DeCapua.

As it turned out, Coronation was a bust. All they managed was three hundred pounds of gray cod. And gray cod was cheap fish. Their catch limit was six thousand pounds of yellow eye with up to 10 percent bycatch. But they couldn't find the yellow eye. They couldn't find anything. Morley started setting gear west of Cape Decision, then off Nation Point, and finally he steamed farther west out to the Hazy Islands. They weren't much, as islands went. There were just three outcroppings of rock ringed by a fifty-fathom shelf. On the gulf side, the shelf dropped off hundreds of fathoms in a few miles. For two days they dumped and hauled back longline gear off the Hazy Islands. It did no good. The yellow eye were not biting.

It was pretty snotty out, though. Gales blew every day and riled the seas. For three days, they set in twenty-foot breakers. It was unpleasant. The boat kept getting caught between high, pointy crests. She went weightless a few times. Some of their lines snarled. Nobody was happy with the sets. They hauled the gear in twisted, the hooks empty and tangled in the line. The wind was bad even in the lee of the islands. Rain cut at their faces like flying carpet tacks. Their eyes burned from all the spray. On deck it was hard to keep their footing, even with a two-inch layer of no-skid padding. With all the seawater and sleet slopping around, it felt as though they were walking on an iron girder wearing skates.

Then the generator acted up. It was on and off like a lightning bug.

They also lost the stove near Coronation. Morley had a roast on when the stack caught fire. He said grease and soot had probably built up in the stack and ignited. Everybody was out on deck hauling and nobody noticed the smoke until the burners were gone. For three days they ate cold ham-and-cheese sandwiches and crackers and bananas. There was no coffee, either. The fire had melted the drip maker. They were able to heat up soup in the microwave. But chicken noodle soup every day, three times a day, drove them nuts.

On the way out they had taken their time, stopping for a hot tub at Baranof Springs. But no one was in the mood for stopping on the way back. It was a bumpy ride. The swells were running at fifteen feet. The *La Conte* rode the swells like a cork in a baby pool, with a lot a babies in it. But nobody got sick.

Just after midnight on Tuesday, five days after it had departed, the *La Conte* limped into Sitka Sound. Morley hadn't paid his harbor dues, so they tied the boat up along the breaker wall between the last cold storage and the fuel pier, just outside of Old Thompsen Harbor.

Morley wanted to leave the following night. The catch had not covered half the cost of groceries and fuel, and they had lost a lot of their good herring bait. But Gig Mork could not go right away—family business, he said. So they stayed in port and untangled a mile of main line, thawed out two blocks of iced herring and busied themselves with little things like cleaning the bilge pumps and filing hooks.

Bob Doyle and Mike DeCapua slept on the boat. It was cold, but there were a few sixes of Bud in the refrigerator to help them shake off the dampness. They watched a video, *The Godfather*, and smoked a joint and lay in bed reading. DeCapua liked reading crime novels. The poetry of violence, he said, was a kick. Bob Doyle read the papers and drank and sometimes went outside to piss over the railing.

Mark Morley slept at his girlfriend's apartment. Gig Mork went on the town. He said he was going out for smokes and that meant he was heading either to Ernie's or the P-Bar or a buddy's house—anywhere he could drink without being hounded. When he was away at sea he did not drink. Otherwise, he was at it all the time. Several people who knew Mork said he'd lost control of his drinking after his older brother, J.R., shot himself to death one night on a cabin cruiser. Other people said that wasn't true. Still, when Mork went out for smokes, he usually didn't get back until he was flat.

"Christ, that Giggy can drink," Mike DeCapua said. He was lying in his bunk, reading. "He's a hell of a deckhand, though. Not many deckhands in his league. And I mean anywhere. That fucker knows what he's doing on a boat."

Bob Doyle said nothing.

"But, *man*, can he drink," DeCapua went on. "He can do that good. He's just a little guy. But he'll drink you right under the goddamned table if you let him."

Bob Doyle sipped his beer.

"I'm having my doubts about the skipper," DeCapua said. "He don't know shit about rockfishing. Maybe he knows about crabbing or pollock fishing. He done most of his fishing out west, right? But he don't know much about rockfishing. I ain't going to quit the guy, though."

"No?"

"Not yet."

"What are you reading up there?"

"I've got a copy of *Guy Claudius.* Ever read it?"

"No."

"I don't do that," DeCapua said. "I don't go quitting a skipper just 'cause he's had a bad trip. I might quit a guy after *three* bad trips. But not after just one."

Bob Doyle said nothing. Having lived years on Coast Guard cutters where privacy and quiet came at a premium, he had learned to let the talkers talk. He would not say a word and after a while the gabbers would gab themselves dry. He stood up. He felt like he needed another beer.

"You hear what I just said?"

"Sure," Bob Doyle said. "Say, are you coming out with us to Whale Park tomorrow? Mark said he might take his girlfriend and her daughter. I'm taking Brendan." That was his nine-year-old boy. Bob Doyle had named his son after the patron saint of sailors. He also had a four-year-old daughter, Katie. But her mother wouldn't allow the girl to go out on the boat. Too dangerous, she had said.

"No, thanks," DeCapua said. He laughed. "You and me and the kiddies watching whales."

"Why not?"

"I'll pass."

"It'll be fun."

It was, too. They were good kids and they loved being out on the water. After breakfast the wind was light and only a few clouds had piled up and the sound was pretty with brassy glints jumping along it. Bob Doyle looked astern, where the wake ran crisply in the calm water. Sometimes Brendan came running up to the railing and pointed out a rising fish. They would watch it drop back in with no splash and the water close smoothly around it. They also watched humpbacks breaching. As one of them came thrusting out of the water a memory flashed through Bob Doyle's mind about the first time he took his family whale watching. Katie had squealed and clapped and her blond pigtails had bounced in such a way that had almost caused Bob Doyle to weep.

Brendan was a dynamo. He reminded Bob Doyle of an otter. At first, he was shy around Mark Morley and his fiancée, Tamara. Perhaps the shyness came from being around Tamara's teenage daughter, Kyla. At one point Morley noticed the boy eyeing the steering wheel. He motioned to him.

"Come on, Captain, get on over here and steer us."

Brendan jumped into Morley's lap, gripped the wheel as though it was about to fly off.

Morley chuckled.

"Relax your grip. Easy now. Okay. That's a little better."

They cut the motors and drifted in the bay. The breeze took away the smell of the whales and lifted the thinly latticed water. The water was dark blue and yet they could see every ridge and wrinkle and sometimes a white, crystalline plume. Brendan and Morley stayed up at the house and Bob Doyle sat on the bow with Tamara and Kyla, drinking coffee and eating tuna-and-onion sandwiches.

They asked him the routine questions, about the Coast Guard, about his divorce. He answered them all as vaguely as possible without being impolite. To steer them off the topic of his marriage he asked Tamara how she had met Mark.

The corners of her mouth went up.

"At the P-Bar"—she laughed—"of all places."

When the clouds had gone a rusty color Mark Morley cranked the engine and headed them back slow across the sound. There was a light breeze and gulls followed them. The water stopped being blue and the mountains went dark except the caps and the town lights shown like a necklace along shore. Near the fuel pier he cut the motors and let her drift up to the break wall. Gig Mork was waiting.

Bob Doyle tossed him the bowline. Mork made her fast to a ladder on the wall. Then he hopped aboard and fixed some buoy floats along her sides. Bob Doyle hugged Tamara and Kyla, and Brendan hugged Mark Morley close and hard, and then they each went up the ladder and stood on the platform.

Brendan said, "Bye, Dad. Thanks for a really, really cool day. Can we do it again soon?"

Bob Doyle hugged the boy and, looking up, saw a car with its lights off and a face through the pane of the car window. He walked with his hand on the boy's shoulder until they reached the car. Then the passenger door opened.

Without looking in, he said, "I'm going to show Katie the boat now, if that's okay."

"Make it quick."

His daughter hopped into his arms, all warm and soapy smelling, and he held her tight and stroked her hair. Then he lifted her up, sat her on his shoulders, took her tiny hands in his and strode off down the slope toward the boat.

He got Katie a soda from the galley refrigerator, put a sailor's cap on her head and showed her around the boat. In the wheelhouse he let her play with the steering wheel. He told her about the boat's engine and its cooling unit and fish holds, and then he put her back on his shoulders and carried her back up the pier to the car.

As he was putting his daughter in the backseat, Morley and Mork walked up. The car door opened and a woman in a jeans jacket stepped out. She had her blond hair tied up in a ponytail and the skin around her eyes was a little swollen.

"Laurie," Bob Doyle said, sweeping a glance past his ex-wife and pointing to the men, "this here is Gig Mork. And this here is my skipper, Mark Morley." She reached for Morley's hand and shook it.

"Nice to meet you," Morley said.

"Likewise."

"Your Katie here is just a darling."

Laurie nodded.

"We have to go," Laurie said. The way she said it made Bob Doyle's throat tighten.

"Something wrong?"

"No."

"So," Bob Doyle asked Laurie, "what do you think of her?"

Morley and Mork looked at each other, then turned and headed back down to the boat.

"You mean that rust bucket?"

"Well," Bob Doyle said, looking off at the harbor, "I suppose she looks worse than she is."

Laurie shook her head. "That thing gives me the creeps."

"We're fixing her up."

"I have to go."

"Give hugs to Brennie for me tonight?"

She opened the car door, got in and started the motor. The headlights came on.

"Thanks for letting him come today," Bob Doyle said. "And for bringing Katie. I'll call you when we reach port."

She was gone.

E L E V E N

That night when Mike DeCapua returned to the boat Bob Doyle was lying facedown on the bottom bunk and looking at the cabin wall. It was dark and the only sound was the lazy creak of the hull in the high tide.

"Sleeping?" DeCapua asked him.

"No."

"Want a beer?"

"I don't think so. Thanks."

DeCapua climbed up in his rack and clicked on the reading lamp.

"How were the whales?"

"Fine."

"Everything all right?"

"I'm just feeling a little low."

"I asked the skipper tonight when we were leaving."

"What did he say?"

"Tomorrow." DeCapua lit a cigarette. "You know what I said to him? I said: 'Why don't we head out on Saturday? We could leave right after midnight.' So he asks me why. And I go: 'It's bad luck to start a fishing trip on a Friday. Don't you know that?' "

"And did he?"

"He does now."

"Then what?"

"So then he goes: 'Cut the superstition crap, Mike.' You believe that? We've been corking off in port all week and now he's worried about one fucking day."

"Well," Bob Doyle said, "he is the skipper."

DeCapua smoked his cigarette.

"So what time are we leaving?" Bob Doyle asked him.

"Four o'clock."

"That's in six hours," Bob Doyle said. "Where did he say we were going?"

"Fairweather Grounds."

The Fairweather Grounds were a cluster of shoals out in the open gulf, about 150 miles northwest of Sitka. The grounds were great fishing, but it took at least eighteen hours just to get out to them.

"What does he want to catch?" Bob Doyle asked.

"Rockfish."

"What's our limit?"

"We're allowed twelve thousand pounds a week, plus another ten percent bycatch," DeCapua said. "I guess that ain't bad. But you know why they extend the season out there, don't you? Because nobody's nuts enough to fish the grounds in the winter. There's fish, all right. But it's hell getting them. Hell. I've fished the grounds. I know. In January, the only things that belong on the grounds are things with fins and gills."

Bob Doyle lay still, listening to the hull creaking around him.

"You ever get spooked at sea, Mike?"

"No."

"Lonely?"

"I been on my own a lot. I figure being alone is the normal course of events. Anything else?"

"You got the one kid, right?"

"No. I got three."

"Three?"

DeCapua coughed. "I got two girls from my first marriage. Haven't seen them for a while."

"How long?"

"Nineteen years."

"That must be rough."

"You got to know how to be alone," DeCapua told him. "It's like knowing how to wipe your ass. No one's going to show you how to do it." He puffed on his cigarette and then said, "What scares you?"

"Oblivion."

Mike DeCapua grunted.

"I don't know. It's just that everything always seems to start off good and turn out bad," Bob Doyle said. "Nothing ever gets any better. And then that's it. You're dead. Know what I mean?"

"No."

"There's nothing that makes you afraid?"

"Sure there is."

"What?"

"Dying with a hundred-dollar bill in my pocket."

"Come on."

"No kidding. Scares the shit out of me. Want to hear a bedtime story? All right. My old man worked his whole life for Pratt Whitney, the airplane-parts company, in Hartford."

"My older brother worked for Pratt Whitney."

"Yeah, well, my old man was a machinist for those fuckers. He went in right out of high school thinking that one day, when he's good and old, he'll have a little something for when he retires. And he worked there for forty years. Forty fucking years. Finally, he retires. Gets a volunteer job with the Boy Scouts. Two months later, he goes on one of them summer camping trips. Cooks for the kids. So now it's Sunday, the trip is over, and he's driving back on Route 5 just south of the Massachusetts line when his heart blows up. Massive coronary. His car rides off the highway. By the time the cops found his old sedan in the ditch, all the salad and hot dogs and shit in the backseat were all stinking and rotting."

"Oh."

"Sleep tight," said Mike DeCapua.

In the night Bob Doyle woke and saw the light in the bunk above him was out and heard the steady breathing and knew that Mike DeCapua was sleeping. He was apparently sleeping well now and not stirring.

Bob Doyle was glad to have his own dry, warm bunk and he was also happy to have some company. He did not want to think about all the times he had neither of those things, and he tried not to think at all. Thinking the night before a trip usually got him in trouble. So as soon as he started picturing somebody doing something he shut his eyes and in his mind froze the person he was giving life to, and then waited for the image he had created to fade.

But after a while he got tired of doing this and his mind began to jump around. Gradually he began to think about Seattle and all of the fiascos, like the time he had gotten so drunk he collapsed in the backyard after he had found out that Laurie had gone to bed with a good friend of his. Then he thought about some of his own infidelities, in particular that leave in Valparaiso, when he and the officers on the icebreaker *Polar Star* had blown a month's pay in three nights at that brothel. That was foolish. But he had done so many foolish things when he was away from home for long stretches that he almost did not take them as seriously as he should have. There was no excuse for his stupidities but the time away from family had not done any good. It hadn't done the kids any good, either.

He thought about that terrible winter when Laurie suffered from post-partum depression after having Brendan. He remembered how she couldn't stop slapping herself and the night he finally had to drive her to the hospital in Newport so she could check herself in for acute depression. Then he thought about that afternoon in Sitka when the hospital called him at work and told him his wife was having her stomach pumped. Later, she insisted it hadn't been a suicide attempt. She had washed down twelve Xanax pills with a can of warm Pepsi so she could go to sleep for a day or two. It was an accidental overdose. Truly.

He did not mind being awake now and he remembered how it had been once when they did not lie to each other and when the lovemaking was sweet and never too hot, or too cold, but steady burning like a well-tended pile of driftwood, and for some reason he began thinking about the afternoon in Newport, Rhode Island, when he first saw her. At the time he was just a third-class SK—a supply clerk—much lankier, with an unruly lack of grace, and short on cigarettes. He went into the exchange and waited on the checkout line behind a woman who was struggling with a bag of limestone. She wasn't strong enough to lift it herself and place it on the counter, so he offered to do it for her. It was not until he thunked the sack on the counter that he noticed the cashier was gorgeous.

"Doesn't he look like such a nice gentleman?" the woman who was buying the limestone said.

The cashier's sassy eyes, transparent as topaz, and full, moist lips threw him into a trance of lust. She looked up from the register keys, glanced at him as if preoccupied and said:

"Looks can be deceiving."

Every afternoon for the next two weeks he bought cigarettes at the exchange. Finally, one night after a pickup basketball game, he walked in determined to find out if she was married, had a boyfriend, or if, God forbid, she preferred women. She was mopping the floors. He fished a bottle of Diet Pepsi out of the cooler, going over every word of his questionnaire in his mind, popped the bottle top and stood, so puppylike it was almost pathetic, in front of the register.

She put the broom down, came over to wipe down the counter and then watched him sip his soda.

"By the way," she said, "my name is Laurie. What's yours?"

He gagged on a mouthful of soda, spitting it out everywhere, and stumbled backward.

"Hey!" she snapped. "What was *that?*"

"I'm sorry."

"I just cleaned that floor!"

"I'm really sorry."

"Here!" She thrust a roll of paper towels at him. "That's not my mess. That's yours. Clean it."

Two weeks later he took her to a Mexican joint, Amigos, and six weeks later he moved into his apartment at the end of a runway where the National Guard launched jet sorties. A year after that they wed at a church in Westminster, Vermont, not far from his hometown of Bellows Falls, in front of two hundred people, most of them his family. She wore an ivory gown and a long train. He wore a tux and kept a flask in his pocket for nerves.

It was the only truly extraordinary thing to happen to him in all of his first twenty-eight years, falling in love with Laura Mae Raymond. And yet, even though it had happened only nine years earlier, it felt as if it were part of his other life, a life that was long gone and buried. History. Not the drinking, of course. But the rest of it. Over and gone and separate, as though someone were telling him the sad story of some other fool on a bar stool at Ernie's Saloon. There was no going back.

Like all of the other things in his life the marriage had started out so promisingly, so lovingly, and then soured. It had started going bad in the second year, after Brendan came along. By the time Katie was born, five years later, they had grown used to wounding each other. Every day they hurt each other a little more without the slightest compunction. Their lovemaking fluctuated between vicious, complacent, inconsolable, maddening, passionless, and then finally infrequent, while the tone of their voices slowly petrified until most everything they said to each other was either repellent, vulgar, stony or, worst of all, indifferent. They were so dead to themselves that even the poisonous talk that accompanies separation—the barrages spouses unleash at each other to make the break permanent—seemed fatigued.

The marriage was actually over long before they shoved off to Alaska. But it was in Sitka that she finally filled the void he had left, or perhaps had never filled, in her heart.

———————

The man she had decided to go down the road with was a flight mechanic
who lived four doors over in Coast Guard housing—an enlisted man, no
less—whose wife had recently dumped him. His name was Rick Koval, and
to Bob Doyle he was a cross between a bulldog and a boar, the size and
shape of a middleweight past his prime. He had a round head set too close
to his shoulders, a bulbous nose above a toothbrush mustache, fleshy jowls,
a small, thin mouth and pig eyes, one of which seemed to wander a bit as
he was talking. The eye had come out of its socket once when he and another mechanic had been horsing around on the hangar deck at Air Station
Clearwater. They were cleaning an H-3 helicopter with long-poled wash
wands, and they jousted and poked each other with them until the other
mechanic jabbed the wand's suction pad into Koval's eye and yanked the
eyeball clean out. The doctors did sew it back in and he regained some vision. But from that day forward the eye had an opaque, faraway stare to it.

Koval was heavy into computer games, hunting, Tabasco sauce, all-
terrain vehicles, keg parties and guns. He was a gunsmith and stored more
than forty weapons in an iron safe at home, except for a German-made
howitzer that fired fifty-millimeter shells. (It didn't fit in the safe.) He
bragged that he'd once shot a buck with the howitzer from a mile off, unfortunately splattering fur and deer meat all over the forest. He often wondered what it felt like to be shot, so he and a buddy arranged to shoot each
other with rubber bullets. They had loaded their ammo, donned several
heavy coats, raised their guns and taken aim when the phone rang. Koval
picked up. It was his mother. Just as he answered, his pal's gun discharged
and Koval doubled over, grabbing his balls. As he wailed and writhed on
the floor, the friend snapped up the phone, said, "Mrs. Koval? Sorry, I just
shot Rick by accident. He'll give you a call back," and hung up.

He might have been a big kid—preoccupied with his toys and replete
with a disdain for authority—but Rick Koval was an exceptional
mechanic. He was also a good welder, the premier machinist at Air Station
Sitka. A chief petty officer once remarked that he was convinced the man
could build an entire car with just a lathe. And so, as small as the Coast
Guard community was, Koval's reputation as a wild man never percolated
up to the command. He raised his share of eyebrows. But he also had his
admirers.

Bob Doyle was incredulous the day he discovered this oversize teenager was his wife's lover. It was the afternoon of his next-door neighbor's garage sale. He was looking at the price tag of an old Venetian-style lamp when he saw Laurie walk up. She muttered a greeting and lit a cigarette.

The back of his neck prickled up.

"Whose lighter is that?" he asked her.

"Oh," she said, and blowing smoke out of the side of her mouth: "Rick just lent it to me."

Later, the woman who was holding the sale pulled him aside. "Listen," she said, "I don't care what happens anymore. I can't stand it. You need to know that she's been having an affair with that guy ever since you got here." It occurred to him that he should feel angry—embarrassed at least—that everyone else on Lifesaver Drive knew his wife was carrying on with a neighbor who occasionally sat at his dinner table. But he only felt sad. He walked home, pulled a Miller Genuine Draft out of the refrigerator and ran a bath. It was blowing outside. He listened to the wind against the windowpanes. He thought he had a fever and felt his forehead. No one was in the house. He got into the tub, shivered and thought he might like a whiskey. All of his important decisions had been figured out while he was having a whiskey. So he got up and went downstairs and fixed himself a Crown Royal.

Days later, when he confronted her, she told him to get out of the house. She won't do it, he thought at the time. Not on my mother's birthday. So he called her bluff. She called him something nasty. He called her something nastier. She called his XO and told him that her husband was beating her. The following day the base commander ordered him out of the house. He threw some clothes and personal items in a plastic trash bag and moved into the barracks. She got an order of protection and, days later, filed for divorce.

Then things got ugly.

Not long after he filed a chit against Rick Koval for committing adultery with the wife of a warrant officer, the flight mechanic stomped up to him in the Eagle's Nest, jabbed a finger in his face and said, "I'm fucking your wife and there's not a damned thing you can do about it." Another day in the smoking lounge, two officers heard Koval's booming voice: "What am I supposed to do to get my life back? Run over him with a truck? Make a grease spot out of him?" Again he reported Koval, this time for making threats against an officer. But the charge was dropped when other witnesses said that Koval was only joking. The flight mechanic was given a warning, though, and ordered to stay away from Mrs. Doyle, at least until she no longer carried the Mrs. in front of her name.

As the summer slid along the scandal turned more sordid and people wound up taking sides. There were those who took the side of the enlisted man, Koval, and thought he was well within his rights to date Laurie. Then there was the camp that thought Bob Doyle had been wronged. It was like some bad play. He began skipping work to follow his wife around town, hoping to catch her and her boyfriend in a lustful act while Koval riled up sentiment in the engineering department against his nemesis. The command, fearing that Koval might walk into the barracks guns blazing, tried to wade into the whole affair. But it only made things worse. Soon people developed opinions about what the command had done and whether it was fair or not fair, and in the end morale only sank even lower.

Then one Sunday, a number of couples who had anchored out in the sound saw Koval, Laurie and another couple returning together from a weekend boat trip. It was all over town within hours and the following morning Bob Doyle filed yet another chit. This one stuck. Koval was court-martialed that September and pleaded guilty to disobeying a written order from his superiors. It came at a bad time for Koval, who had just put in for a promotion to warrant officer; he was busted down a pay grade, though ultimately the court did allow him to stay in the Coast Guard.

The hobbling of Koval did nothing to prevent Bob Doyle from walking his own gangplank. He drank day and night, Miller Genuine Draft by day, Crown Royal by night, mostly because they were convenient. When he was transferred to Sitka and promoted to supply officer, his immediate supervisor, a lieutenant by the name of Bill Adickes, put him in charge of overseeing the exchange and the Eagle's Nest, a pub for officers complete with outdated jukebox, pool tables worn to the threads and cheap ashtrays. It was, in retrospect, almost funny how the Coast Guard could put an alcoholic in charge of all of the booze and cigarettes he could possibly want. He took full advantage. Before closing the exchange he'd nick a couple of cases of Miller off the front display stack, a few packs of Luckies, a bottle of Crown Royal or Canadian Hunter for good measure, and polish them off at home on his couch.

He would skip work, or show up, frazzled, after lunch. His subordinates covered for him for months. But the exchange manager, one of the only people he'd ever pushed to do her job and who thoroughly despised him, went to Adickes, his supervisor, and ratted him out. One day Adickes confronted him about the missing merchandise and he denied taking it.

"Your Coast Guard career is over!" Adickes shrieked. Whenever he got upset, the veins in his neck looked like they were about to explode. "I'll make sure of that. Now get your stuff and get out of my office!"

Bob Doyle had not gotten angry or violent over that, either. Why hadn't he? Most anyone else would have fought back. Defended his honor. Maybe he had no honor anymore. Maybe he never should have become an officer. The Coast Guard policy had always been to move up or ship out. He liked being a Coastie, so he'd had to rise through the ranks to stay. But he never fit as an officer. It wasn't just a simple progression. He didn't like the politics of officialdom. The responsibility. Maybe that's where the Coast Guard failed him—by promoting him.

It all came to a head two days before Thanksgiving in 1996, when the cops pulled him over into the parking lot of the Mormon church on Sawmill Creek Road. He'd downed more than a few boilermakers at the Eagle's Nest and had been soaring around town in his van. The charge was driving under the influence, and they locked him up for three days.

His conviction came down in January. It was his second offense in a year, and by the book, the Coast Guard had every right to throw him out. Adickes suggested a graceful exit—an administrative discharge, he called it. And so, on April 30, 1997, twenty-one years, five months and twenty-seven days after joining up in Springfield, Massachusetts, Bob Doyle signed the forms for early retirement.

It was better that way. The divorce had already gone through, in January, and he'd had to endure seeing Laurie and Koval together on the chow line in the mess. He tried to keep with his children. The custody settlement allowed him to see Brendan, who at that time was eight, and Katie, three, on weekends. Since the court allotted his ex-wife the family van, he would send a taxi over to his old house to pick up the kids. He could never bring himself to see them walk out of another man's house.

By then he was homeless. On weekends he rented a room at the Super 8 motel so the kids would have a place to sleep when they came to visit.

Late that summer he heard that Koval and Laurie had been granted permission to live together with the kids in Coast Guard housing. Then there was talk that Koval was being transferred to Kodiak in the spring of 1998, and that they planned to marry before the move.

That was just three months away, now. He wouldn't miss those two. But the kids. They were going to take his kids. Brendan and Katie. Unless he made money soon, enough to get himself off the streets, the state of Alaska was going to wink and nod and let them get away with it.

The boat did not rock as much as it had earlier and gradually Bob Doyle stopped thinking about his ex-wife. Thinking about her won't do you any good, he said to himself. God, I hope we have a good trip and make money. It would be wonderful if I could stand up and do something right for a change. Maybe I will, he thought. And then he was asleep.

T W E L V E

When he woke Bob Doyle noticed two, big feet dangling off the edge of the top bunk. The toes were black and blistered, the heels scaly and scabby and puckered. They smelled like rotting cabbage. Bob Doyle tried not to breathe through his nose while he dressed. He pulled on a pair of sweats, a hooded jacket, a wool sweater, socks, rubber boots and a cap, and then stood for moment, contemplating the big, gnarled feet. He was going to have to do something about Mike DeCapua's feet. Maybe boric acid and a wire brush would help. He stumbled into the galley.

Pouring a cup of coffee, black, he heard voices on the dock. He pulled the Dutch door open, went out to the starboard railing and looked at the channel. It had rained and there was a mist. He loosened his sweats and pissed off the railing, watching his urine make a long, steaming arc, then shook himself and, wiping his right hand on his sweatpants, walked around to the foredeck. Mark Morley and Gig Mork were bringing aboard the last of the groceries. Morley tossed a duffel bag on the deck along with a small, vinyl suitcase. Mork had a bulging, plastic trash bag over one shoulder.

"What's in the bag?" Bob Doyle asked him.

"Smokes."

"Hey, Bob," Morley said, "give Giggy here a hand. Let's get all of his shit in the fridge and cupboards quick. I don't want to fart around here any longer. Where's your buddy?"

"Sleeping."

"Well, get his ass up. We got a trip to make."

"Right."

After he and Mork began packing up the refrigerator and freezer and stowing the tins, Mike DeCapua came out. He was in his boots and rain jacket and powder blue pajama bottoms with squiggly circle designs. From his cracked lower lip balanced a bent cigarette, as though it had been glued there.

"Hey, Giggy," he said, the cigarette jiggling like a doll on a coiled spring. "Give me a light."

Mork threw him a lighter.

"Lot of food."

"Give us a hand, Mike."

"In a minute."

He went out on deck and smoked while they packed the groceries away. There was maybe two weeks' worth of food, Bob Doyle thought, and it took them an hour to squeeze it all in. In a little while they heard the cough and grunt of an engine turning over, and then the throbbing through the deck planks, and everything had that sooty, diesel smell. Morley had grabbed DeCapua and together they were testing the motors, bilges, deck lights, winch, RSW lines, fuel tanks, condenser, fuel filters, injectors, air intakes and batteries. In the meantime Mork showed Bob Doyle the gear in the bait shed, the totes in the holds and the grungy, snarled line in buckets on the roof of the pilothouse. They were going to have to clean and sharpen all of the big hooks and splice line and check all of the leaders and swivels.

He started off in the bait shed, rearranging the blocks of iced herring and chum bait and tidying up gear. Then he heard the skipper and Mork talking out on the dock.

"This better be a good trip," Morley said.

"It will be."

"If it wasn't for that bastard I'd be a lot calmer about everything. Every ten minutes that cell phone of mine rings. What the hell does he do that for?"

"You know owners."

"But that goddamned phone rings all the time. Swear to God I'm going to toss the fucker overboard."

"Take it easy," Mork said.

"You believe he wouldn't buy us any more bait? Who does he think I am? The fucking bank of Sitka? Christ. I had to use my own credit card at the cold storage for the last of it."

"We'll use the chums."

"How's that old gear?"

"We'll get it into working shape on the way out."

"Good."

Later, while Bob Doyle was carrying line out on the foredeck, Morley called to him.

"Hey, Bob," he said. "I want you to meet somebody. This here is David Hanlon. He'll be fishing with us. David, this is Bob Doyle, my newest deckhand."

Bob Doyle lifted a hand and the stranger took it, softly. A breeze hit the man's face high up, lifting his fine, black hair around his ears. "Nice to meet you," Bob Doyle said.

The man only nodded.

"You'll be sharing the stateroom with Giggy. It's a big old room," Morley said. "Bob, take David here inside and show him around. Get him a coffee or something. You want some coffee?"

"All right."

"Here, let me get your bag," Bob Doyle said, bending over and grabbing the straps. "I got a pot on inside. Come on."

Bob Doyle had known some Natives. He'd met them in the bars, along the waterfront. The younger ones drank a lot and tended to jabber. The older ones were more cryptic, gloomy. They'd look through you, not at you, and speak in the deep, somber tones of the vanquished. This Hanlon guy wasn't like that, exactly. He was quiet as a tall glass of water, and when he did speak his voice was soft and dry, like the rustle of well-worn leather. The glasses make him look like some graduate student, Bob Doyle thought.

His eyes were deep-set eyes, black as a sparrow's, untouchable. He had thick lips, a broad nose. He wore a tight, faded fleece jacket and Bob Doyle had seen the muscles of his shoulders twitch under it when he climbed over the gunwale. He's strong, Bob Doyle thought. Don't let that shy, timid stuff fool you. The man is a bull.

He showed Hanlon to his bunk, then took him to the galley and poured two cups of coffee. Mork, then DeCapua, came in. Mork had a half frown on his face. They all shook hands and then Morley called Hanlon down to the engine room.

When they had left Mork said, "That guy is a spy."

"How's that?" DeCapua said.

"A spy. The owner's hired gun. He's here to watch us and give the owner an earful."

"Why?" Bob Doyle asked.

"He ain't gonna sit on his ass and give orders, is he?" DeCapua asked.

"No," Mork said. "He'll be working all right." He poured himself a coffee and gulped it. "I'll find him plenty to do."

"Is he any good?"

"Skipper says he knows how to fish and can put on a bait right quick. But that ain't what bothers me."

"What then?"

Mork lit another cigarette and went out and climbed up the steel ladder to the pilothouse. They could hear his boots going back and forth through the ceiling. Bob Doyle looked at DeCapua and shrugged his shoulders.

Morley waved at Mork to toss the stern line on board and then kicked the starter. The boat sprang off the wall. He throttled up the engine until it stopped grumbling and soon the fuel pier was behind them and they were cruising through the low-wake zone and on down the channel.

They fast-coiled the dock lines, stowed them in the lazarette, then lifted five fifty-pound boxes of iced bait from the forward hold, carried them astern and lashed them to the side of the bait shed. Mork told them to make sure every five-gallon gas jug and buoy ball was tight on the railing or they would pay for any lost gear. Then he turned and climbed up to the pilothouse. DeCapua and Hanlon went to their bunks. Bob Doyle stood alone on the stern.

Along Halibut Point Road he saw two spokes of yellow light, probably those of a truck, and heard the tearing hiss of tires on wet pavement. He looked up. Dipping and slicing through the air, a seagull was following them. It kept its distance, eyeing their wake. He lit a cigarette. Opportunists, he thought. They were passing the jetty now. He felt the smooth chug of the engine, opened up to two knots. Soon they would be at Salisbury Sound. Then they would head outside, steam a mile west of the Sisters, and then, near Calves Head, crank the boat up to six knots and run a slot north and west. There was plenty of water between them and the Fairweather Grounds. With luck, they'd be there in eighteen hours—ten o'clock that night. Mark had said they would stand four-hour watches on the way out.

Bob Doyle was thinking that three months was a long time to go without a payday. What did they call it? Being in hock. That was it. He was in hock, all right. The Sitka fishing fleet was in hock, too. The whole damned country was in hock, come to think of it. Everybody was a member of the Hock Club. But three months was too long to go without a check. His monthly retirement check. What a crock. Work your whole life thinking you'll get a pension and then everyone else takes a bite out of it. Maybe Mike was right about pensions. Fuck pensions. The city of Sitka was taking out of his pension for the jail time he did and for court costs. The Coast Guard was taking out for those checks he'd bounced at the Eagle's Nest and for renovating the house they'd kicked him out of. His ex-wife was taking out for child support. Once everybody else got done with his check, there wasn't enough left over to stuff a clam.

He wondered if the weather would hold. If they caught their limit he might even get a few bucks, once they had covered the cost of the groceries, bait and fuel. If he didn't fuck up, Mark might even take him along tendering in February. He could count on a hundred bucks a day doing that. Guaranteed. No worrying about the size of the catch or the weather or the price of fish when they reached port. One hundred bucks a day. That was decent money. Do that for thirty days and you might be able to rent a place of your own, he said to himself. Some small, quiet place, out the road. The kids could sleep in his bed. He'd stay on the couch. Wouldn't that be something.

They were coming up on the Old Sitka Rocks now. There was a sudden knot of lights on the shore. He picked out one, a faint yellow circle. That's where his Brendan and Katie were. Right there, at that light. It was a streetlamp out front of a duplex with brown siding. Real quaint. He'd seen it once from the street and then another time, New Year's Day it was, while sailing out on the *Min E*. The boat was slipping along the shore and he'd hurried to get the binoculars and had even spotted Brendan and Katie playing in the front yard. He had wanted to shout to them. But he had not been able to get his jaw to move.

They wouldn't have heard him anyway.

Standing there, his gaze frozen, blank, as though he were staring at a burning house, he pictured his children sleeping. Katie liked to sleep on her side, Brendan on his stomach. He wondered if sleeping children could sense when their parents were thinking of them—have an unconscious thought, a sixth-sense sort of feeling that would wake them and make them say, *Hey, Dad was just in my dreams*.

Oh, cut the crap, he said to himself. You drink and carry on and spend half your kids' lives away from home on icebreakers and you lie and lie and lie and you nick stuff from your employer and then you lose your wife and you want to get all soppy about it now? Get some sleep. You're going to need it. You won't sleep again for two days.

They were passing the ferry terminal. The tops of the spruce picked off the moonlight. Gulls called and waves broke on the rocks. He felt on the edge of something. He blew on his hands and turned to go inside.

"Please look over my kids," he said out loud, to no one in particular.

No one was up before ten. The seas were flat and there was only a light breeze. They ate hot oatmeal and scrambled-egg sandwiches on toasted bagels with ketchup and cream cheese and everyone had several cups of coffee. Bob Doyle took his time eating. He felt sleepy and he felt good. The mountains had dropped below the horizon behind them and the gulf was rippling out ahead in sharp, minute splashes. The water had deepened in color but the breeze and the sun spackled the wave tops white. The crew seemed upbeat. Everywhere they went on ship there was the close, soothing hum of the engine, the sharp, dry air of morning. And they were back on open sea. They were on it and they were a part of it, and as the bow smoothly parted the chop it felt as though a world of impossible changes was peeling open before them.

They had fifty skates of gear to get ready. That was roughly ten miles of three-quarter-inch monofilament line. With the new line they had, they could make thirty skates. With the fairly decent gear, they could make another ten. Then there was the old stuff. This line was frayed from rubbing on sea rocks. The hooks that went with it were rusty, the tips crusted with rotted herring. It was snarled and tangled and twisted up, and most other skippers might have just chucked it and bought new gear. But Mark Morley thought it could be retooled. With a little patience, a little sweat, maybe some of it could be made useful again. Maybe all of it.

Bob Doyle crawled up the side ladder and lowered the buckets to Mork from the wheelhouse roof. They lugged the buckets down the gangways to the bait shed.

Mike DeCapua barked, "What's all that?" Bob Doyle had just set down a bucket of the snarled line on the shed floor.

"Gear."

"Wrong," DeCapua told him. "It's shit gear." He reached into a bucket. "Oh, fuck. Look at this."

The line had been used to fish halibut. The hooks were eighteen feet apart. For yellow eye they would have to tie on new gangions and hooks and space them apart every eighteen inches. It was pretty frazzled line.

"Who told you to get this?"

"Mark."

"He still up in the house?"

"Yup."

DeCapua stomped out. Bob Doyle found a pair of rubber gloves, turned over a five-gallon bucket and sat down on it. He glanced over at David Hanlon. Hanlon was sitting hunched over and quiet, altogether within himself, not breaking eye contact with what he had in his big hands. Bob Doyle watched those hands. They were making clover-shaped loops in the line. They were tying gangions, setting snaps, impaling herring on hooks, smoothly, and then salting each strip of bait. The fingers moved lightly, slipping up and down the line, gliding in and out, as if they were playing harp strings. The fingers opened and closed, opened and closed, moving as they must have moved thousands of times before now, in a blur, in a rhythm, in their own time. Loop, herring, salt. Loop, herring, salt. Loop, herring, salt.

"That's good baiting," Mork said to Hanlon. Mork had just come down from the pilothouse. Hanlon did not look up.

"How about I give you guys a hand," Mork said.

"Sure," Bob Doyle said.

The three of them built skates. The herring they had was not fresh. Fresh herring was easier to impale on the hook. Maybe that's why it's more expensive, Bob Doyle thought. This bait had been frozen and thawed, probably more than a few times. But it was all right. A yellow eye would snap at it if a hook came down and smacked the fish on the snout. The boat swayed back and forth, and they kept on working. Every so often Bob Doyle would jab his finger on a hook or catch his rubber glove in a stainless-steel snap. Hanlon worked without saying anything. The tips of his fingers went white as snow. But they kept gliding along on the gear.

"What's the matter?" Mork asked Bob Doyle.

"Ah, it's my hands. I'm cutting them up good."

Mork went on baiting. "Tonight, after we're done, go out on deck and take a nice, long piss on them."

"What?"

"That's right," Mork said. "Fish are full of bacteria and shit. You got any cuts, they'll get infected. Poisoned. So make sure to piss on your hands later on. Cleans 'em out good."

"I will," Bob Doyle said.

Outside it was getting cold. They turned on a light and could see their breath in it. Bob Doyle worked awhile, baiting, until the sweat on his forehead began to chill. The wind whipped at the corners of the shed. A draft was coming under the door and down the walls.

The door swung open and an icy wind rushed in, followed by the skipper's head. Bob Doyle saw it shadowed against the dark. "How about some sound, fellas?" Morley asked.

"How about some heat?" Mork said.

"Got no heat."

He drilled three small holes through the aluminum walls behind the door, poked some copper wire through them and twisted the wire around the contacts of two speakers. Then he hung the speakers from nails.

"You guys like the Stones?"

"They're all right."

Morley went back out and a few moments later Mike DeCapua returned. He pulled up a bucket, plopped himself down and took up a line. The album *Let It Bleed* was playing. The speakers sounded a little tinny, but the sound was coming through clear enough. DeCapua started whistling. He liked to whistle. He whistled louder when "You Can't Always Get What You Want" came on. Before long he was crooning the refrains, humming the lyrics he didn't know and keeping time with the drums by bobbing his head and tapping a steel hook on the side of the bucket.

When the song ended DeCapua jumped up and threw a handful of herring on the floor.

"Fuck this."

"Where you think you're going?" Mork asked him.

"Coffee."

The door banged shut. Bob Doyle looked at Hanlon. He had not missed a beat. His hands were still playing that invisible harp.

By two-thirty they had thirty skates done. The herring had run out and they were baiting now with chums. They had a lot of it. Every so often Mork

would go up to the pilothouse to update Morley and check their position
on the charts while the rest of them baited. Mike DeCapua was telling stories of coke parties, his days as a train hobo, how he camped in the "jun-
gles" near the rail stops and made friends with brakemen by sneaking them
dope, and how he and the rest of the "Bad, Bad Boys from Dog Patch" at
the ANB Harbor once pooled fifty bucks and bought a 1948 aircraft hangar
and converted it into a floating no-tell motel. Bob Doyle kept sticking his
fingers with hooks. When he cursed the gear he got a rise out of everybody,
even a faint smile from Hanlon. Mork would offer the others cigarettes and
bring back mugs of coffee. It felt good to hold a warm mug.

At one point Mike DeCapua said, "Hey, Giggy, what's with the
chums?"

"What?"

"Ain't there any more herring?"

"No."

"Beautiful."

Mork scraped a hook with a file.

"Fucking hands," DeCapua said. Mork looked at him and went on
scraping. DeCapua opened and closed his fingers, slowly. "My hands are
fucking numb."

"Want me to piss on them for you?"

The others snickered.

"How many skates we got left?" DeCapua asked.

"Well," Mork said, "there's all this here and that old stuff in the
buckets."

"I ain't doing that."

"Hell you ain't."

"Hell I am."

Mork jabbed the hook through the chum's mouth and out its gills.

"Hell you ain't," he said quietly.

DeCapua looked down into a bucket. "Who the fuck would do this to
longline?"

Mork pulled a pack of Marlboros out of his shirt pocket.

"Have a cig, Mike?"

"What's that?"

"A cigarette. Want one?"

"Yeah, I'll take one."

Mork drew a cigarette from the pack.

"Need a light?"

"Sure. Gee, thanks, Giggy."

Mork held out his lighter, flicked it and waited until DeCapua took a puff.

"Now," Mork said, "shut your fucking yap."

They smoked and baited. Mork offered Hanlon a cigarette, but he said no thanks. He used to smoke, he said, but not anymore. He used to drink, too. For a long time he was a rummy. A whiskey rummy, a beer rummy, a rum rummy, a pretty much anything rummy. But he did not drink anymore, he said. Not for six months. Going on seven.

"How did you stop?" Bob Doyle asked him.

Hanlon looked up from his gear. "I don't know," he said. He shrugged his big shoulders. "I just stopped."

"Did you go somewhere?"

"A place in Sitka," Hanlon said.

"And how was it?"

"Not bad."

"But how did you stop?"

"I don't know," Hanlon said. "I wanted to, I guess."

Mork asked him where he had fished and Hanlon told him pretty much everywhere. He was from Hoonah. His father was a Tlingit leader there, of the Shark Clan. He was also an established commercial fisherman, with several large vessels. One of them was the *Claudia H*, a fifty-eight-foot seiner. His father was a proud man. He took him out to fish for the first time when he was thirteen. His older brothers had been fishing since they were six.

"Why did he wait till you were thirteen to take you out?" Mork asked him.

"I was sick a lot," Hanlon said.

"What did you have?"

"Anemia."

He told them his father and brothers had fished everywhere around Glacier Bay. Up and down Chicagof Island, the Icy Strait, all the way up to Yakutat and the Fairweather Grounds.

"He was a great fisherman," Hanlon said. "He knew the tides. That's what it's all about. Knowing the tides. You know the tides you catch the fish." He looked up. "That's what my father always said."

Bob Doyle said, "Where's your father now?"

"Dead."

They went on working.

"So how'd you meet the owner?" Mork asked him. Hanlon stuck a hook through a bait.

"In Sitka."

"Yeah?"

"During my rehab."

"Oh, what the hell, Giggy," DeCapua said. "Why don't you just ask him why he's really here?"

"What?"

"You want me to ask him?"

"What the fuck you talking about?"

"You don't know, huh," DeCapua said. "How about the skipper's cell phone, for one thing? It rings all the time. Whenever we're in range that thing's ringing. That's the owner calling, isn't it?"

He turned to Bob Doyle.

"You just go up to the wheelhouse?"

"Yeah."

"Hear the phone ringing?"

"Yeah."

"Didn't answer it, did he?"

Bob Doyle looked away.

"See?" DeCapua said. "He don't answer his phone. Told me not to answer it, too. Why? Huh? Why don't he want to talk with the owner?"

"That's enough," Mork said.

"What?"

"Shut up."

"We're fixing shitty gear and baiting shitty chums. Why are we fishing with chums? They're cheap. Owner gave them to us for nothing. But we aren't talking with the owner. Why not?"

Mork dropped his skate. He squared his shoulders and fixed his dark red eyes on DeCapua. His lips were working.

"Oh shit," DeCapua said. "Mom's mad."

"What you say?"

"Nothing."

"You say another word I'll kick your ass."

DeCapua looked down at his skate and whistled. Bob Doyle waited for it then. But nothing happened. Mork just sat there on his bucket with his face twisted in disgust. DeCapua put his hands on his knees and stood up.

"Hell," he said. "It's time for my watch."

"Is he always like that?" Mork asked. DeCapua had already gone out.

"He's a talker," Bob Doyle said.

"He's a pain in the ass."

"Sometimes."

"All the time."

"Don't let him get to you," Bob Doyle said. "He likes to talk. Most times he talks for the sake of talking. He's not so bad. And he's good at what he does."

"He gets on my nerves."

"Well, don't let him."

With his lips, Mork drew another cigarette out of the pack, lit it and let a curl of smoke slip between his teeth. He said, "Let's just get our twelve thousand pounds and get the fuck home."

Bob Doyle nodded.

"That fucking asshole," Mork said.

Then he reached down and picked up another line.

F O U R T E E N

They had finished a supper of casserole and soup and Marlon Brando had just appeared in *The Godfather* when Mike DeCapua banged down the ladder to the galley and told everyone he was done with his watch. Bob Doyle picked up a pack of cigarettes, fixed himself a mug of coffee and climbed up into the wheelhouse. He looked into the partition behind the skipper's chair. Mark Morley was lying on his side, snoring. Bob Doyle sat down in the chair and looked out to sea.

The boat was on autopilot, so he sat watching the wheel move itself. It was a strange thing to be sitting at the helm of a ship that did not need to be guided through the night. He sipped his coffee and smoked. There were a few stars out. He could not see the swells but he could feel the slow lift of the boat and the soft, heavy settling of the stern when one passed underneath. The *La Conte* was taking the swells on her quarter, settling nicely as the swells rolled by, and he sat in the high chair and looked for a light, something to break the blackness.

After a half hour he saw a flicker of light, like a firefly in the woods on a muggy night, to starboard. It was such a small point of light that he wondered if it was a ship at all. In a few minutes he saw it was a fishing boat, and as the ship grew closer he saw that its outriggers were up. They're heading in, he thought. They're running a slot to Cross Sound. I wonder if they filled their totes. I wonder how many totes they have. I wonder if they were out on the Fairweather Grounds. I wonder who is waiting for them at home.

Every now and then he would glance over at the computer. On the laptop you could monitor the *La Conte*'s track. It showed that the ship was following a true course, driving just east of the thousand-fathom line of the Alsek Valley along a four-hundred-fathom gradient. Where they had sailed appeared as a solid black line. Where they were going appeared as a hash line. The seafloor showed in reds and browns and blues. That's a whole other universe down there, he thought. It was a world of canyons and plains and mountains and deserts. He tried to picture how it would be down where the rockfish were swimming. It would be quiet, of course, and dark. It would be very cold, too, he thought.

He noticed a gray silhouette on the screen, and beneath it the words *Submerged Wreck*. The gulf was peppered with explosive dumping areas, cables, sunken destroyers and cruise ships. This one had the shape of a fishing vessel. He wondered how long ago it had sunk. The vessel looked like it had turned on its side. There was the outline of her tail and her propellers, stuck in the mud. He imagined a boat covered in barnacles and coral and waving plants, a dark, watery tomb.

He settled back in the seat and sipped his coffee and lit another cigarette. He heard a cough and watched Mark Morley shuffle out the side door and climb down the outside ladder. Within a few minutes he heard boots on the stairs. He glanced at the video plotter and rubbed his eyes.

Behind him, he heard Morley say, "So, how's it looking, Captain Bob?"

"A boat passed us a half hour ago. Looks like we're going to be alone out here."

"When was the last time you checked the bilges?"

"An hour ago. They were just a few inches up. I pumped them out."

"Good."

They stared out at the perfect blankness ahead.

Morley said, "I really hope this works."

"Me, too."

"I really mean it. I hope this trip works for all of us."

"I do, too," Bob Doyle said. "I really do. And I hope you keep me in mind for that tendering job in February. I'd like to work with you on that."

"I know."

"Hey," Bob Doyle said, "that was a lot of fun yesterday out at Whale Park."

"It was, wasn't it?"

Morley removed his glasses. When he did Bob Doyle could see the dark pouches under his eyes that the thick lenses had hidden. He thought the skipper's eyes wandered a little, too, almost like a blind man's eyes.

"So," Morley said. He put his glasses back on. "What really happened?"

"With what?"

"Your wife."

"Oh," Bob Doyle said. "She had an affair and we split up. It was an enlisted man. A flight mechanic on the base."

"Feel like talking about it?"

"Not really."

"All right."

"There's not much to tell," Bob Doyle said. "She went for this big, twitchy fucker who collects guns and plays around on the hangar decks with laser-dot scopes."

"Damn."

"One time the prick shot himself by accident through the hand with his own forty-five automatic."

"Wow."

"He wanted me to take a poke at him so he could court-martial me. I was an officer and officers can't pick fights. You lose everything if you do."

"You didn't hit him, did you?"

"No."

"That's good. Violence doesn't get you anywhere. One minute you think you can do some good kicking somebody's ass, next thing you know you're worse than the thing you're fighting. Take it from one who knows."

There was a card wedged in the glass of the lookout window. It was one of those miniportraits of Jesus that Bible salesmen hand out at supermarkets. Bob Doyle nodded at the photograph of the young woman beside it.

"You always bring her along with you?"

Morley nodded. "Tamara's a good girl. We've had our ups and downs. But she's my world."

Bob Doyle pondered the photo.

"I was a lot wilder in my younger days," Morley said. "Drank a lot. Never saw a lady I didn't like. Well, maybe a couple." He laughed, then said, "I've been in trouble with the law, you know."

"Really."

"No shit. I did time for a hitting a cop. Not a lot of time, mind you."

"No?"

Truth was, he had served fifteen months at a correctional facility in Anchorage for kicking in a cop's face in Valdez. That episode was the grace note to an agony-to-ecstasy-to-agony tale that had actually begun two years prior to his arrest, on a rainy summer night in Sitka.

That night, Mark Morley had been chugging Jim Beam inside the Pioneer Bar and popping off challengers at the pool table when a little guy with a New York accent and a nifty parka chalked his stick and put a bet on the bar. At first glance, he didn't look like a pool-hall hound. He looked like a chump. But he wiped the table. That impressed Morley. The man had a college degree. And he was an Alaska skipper. To Morley, a high school dropout from Detroit who was the son of a West Virginia coal miner, polish and sea savvy were qualities to hold in high regard.

They bought each other whiskies. Rob Carrs introduced himself and in no time Morley was pouring out his woes. He had just bailed from a

Bering Sea trawler and come to Sitka to find work on a salmon troller. Literally, he had missed the boats; everybody was already crewed up for the season. He did wangle a job trolling on the *Tiffany* with a bunch of Mexicans. But the crew never made a dime. The owner always stiffed them.

Morley was too proud—and hungry—to go on like that. He'd packed his bags and was prepared to operate forklifts in Detroit for the rest of his days. Relax, Carrs told him. Alaska had its share of fool's gold. But there was plenty of real gold to be had, too.

The next day Carrs introduced him to Dave Franklin, owner of the *Heida Warrior*, the troller he'd been skippering. Morley was in luck: the boat had just lost a deckhand. Carrs vouched for him and by that evening the boat left Sitka with Mark Morley in tow.

The *Heida Warrior* fished the last opening of the season, unloaded in Ketchikan and returned to its port of call, Sitka. The trip was a success. In Deep Inlet, they caught a heap of dog salmon and sold them for forty thousand dollars. Everyone, including Morley, pulled a nifty share. That was the end of any thoughts of returning to Michigan. The man was hooked.

The following season they did even better—until Morley got into trouble with a barmaid at the Marine, a strip joint in Ketchikan. She caught his eye with her overly tight Bush Company T-shirt. He caught hers with his tangerine-colored sport jacket and a line delivered with just enough excess of Southern drawl: "Hey, *dawrlin'*! What's a sleazy girl like you doing in a nice place like this?" She rolled her eyes haughtily and sauntered off. He howled.

A few hours later she returned in a body suit. They and Carrs had a few more drinks before stumbling next door to the Frontier Bar. In a booth, Morley took off his glasses, and they started squeezing. She stuck her tongue in his ear and offered him a ride home. Carrs took his cue and waved to the couple on his way out the door. They hardly noticed.

The next time Carrs saw his deckhand was at three the following afternoon at the local jail. He found Morley slumped in his cell, still in his tangerine jacket, looking as glum as a lost beagle. His lovely lady had apparently filed rape charges. According to Morley, they had driven to a parking lot at the ferry terminal. She offered a blow job. He accepted. Because he'd had much to drink it was not easy going, and she sat up, annoyed, and said that on second thought her husband might not understand her tardiness. She shoved him out of her car, the job half done, as it were, and he was walking back to the wharf when a police car rolled up.

Unfortunately for him and the crew of the *Heida Warrior*, the judge did not buy Morley's version of events and ordered him held for two days until his bail hearing. Dave Franklin had to post ten thousand dollars bail, Carrs had to take formal custody of his deckhand and a trial date was set nine months hence.

When the *State of Alaska* v. *Mark R. Morley* came to trial, everyone turned out for the show except the plaintiff. Because of her unexplained absence, the prosecutor had to rely on the statement she had given police the night of the alleged incident, in which she said she could not recall having any sexual contact with the defendant. Normally, the case would have been thrown out right away. But this one had been advertised in the newspapers for months, so they took the proceeding to its hazy conclusion.

The verdict came back not guilty, but it was obvious that Morley was shaken up. He had lost his beard, locks and ponytail—he had to snip them for trial—and much of his swagger. It took him months to get over the experience. He quit clubbing, topless dancers, pot parties. He worked all the time and, as a token of gratitude to Dave Franklin, built a bait shed on the aft deck of the *Heida Warrior*. What Morley lacked in sea smarts and tact he compensated for with enthusiasm and loyalty. He did not snivel, gave 100 percent effort and was always ready to sail on an hour's notice. His hard work paid off. Franklin and Carrs rewarded him with a full-time job as deckhand.

By the fall of 1994 Morley had paid off his debts, settled up with the IRS and had twenty thousand dollars of his seining share stuffed in his duffel bag. He'd had no time to spend money; the *Heida Warrior* fished without letup. In mid-September, after a one-day stop in Sitka to pick up bait and longlining gear, they took off for Cape St. Elias and a forty-eight-hour halibut opening that Dave Franklin did not want to miss.

He wasn't after the halibut. Alaska's fishing authorities had forgotten to set a limit on what boats could take as black-cod bycatch. At the time, halibut was selling for fifty cents a pound; black cod for three dollars a pound. Plus, Franklin knew a hot black-cod hole near Cape St. Elias. It was a no-brainer. They sailed straight to it and dumped every skate they had, 150 all told. They were still fishing the hole when a storm snuck up from the southwest and caught them with 70 skates still in the water.

A gale was blowing and the seas were up to thirty-five feet when the cooling unit blew. The catch began to spoil. Most skippers probably would have pulled their gear, turned south and clawed back to Sitka against the blow. Not Carrs. He ordered the crew to leave the gear in the water. They made straight for the nearest port, Cordova Bay, the longlines trailing behind.

It was a wild ride but a quick one. Instead of eighteen hours, it took them twelve to reach Cordova Bay. They anchored and started hauling back. They were stunned: every single hook had a huge black cod dangling from it. The total haul was seventy-five thousand pounds. After deducting for groceries and supplies, each crewman walked away with twenty thousand dollars—not bad for two days work. And more was to come. No one in the fleet, it turned out, had gone for halibut. Another opening had been scheduled in a week's time. Franklin parked the *Heida Warrior* in Valdez, told the crew to be ready to go when the weather cleared and hopped a plane with Carrs to Sitka, leaving Morley behind to mind the boat.

It was Mark Morley's finest hour: he had weathered a humiliating trial, gained his sea legs and the respect of his mates and, with forty thousand dollars in his duffel bag, was the richest he had ever been. Not only did Morley have the dough, he was in an Alaska boomtown flush with swank restaurants and 24/7 cabarets. The tin man from Motor City had made it to his Emerald City.

He decided to celebrate his good fortune; really celebrate. He drank, smoked and snorted his way to dizzying heights, along the way wooing women, his Achilles' heel. Valdez had no shortage of female flesh: divor-cées, single moms, ladies of means with a misstep or two in their lives, waitresses, strippers, lap dancers, barflies—they were all hard-partying gals, a little rough around the edges, and all lookers. Every one could see Morley coming with his Detroit street strut, his California ponytail and nothing but greenbacks and time on his hands.

When Carrs returned a week later, he found his deckhand had blown nearly ten thousand dollars of his earnings. Morley had girlfriends in every joint in town, though there was one tart, Lisa, who had hooked her float to his parade and refused to let go. She was a runaway from the Midwest, early twenties, a little goofy—she claimed she saw angels walking the streets—with a stringy, bleach appeal. She'd had a few run-ins with the local police, and was not exactly a hit in the clubs. (One bar manager had eighty-sixed her for soliciting on the premises.) It was Lisa he took to the Sugar Loaf dance club the night of October 2, 1994. The bouncer told her to scram. She refused. So the bouncer called the cops. While the couple danced on the floor, two officers entered, grabbed the young lady by the arm and led her out to the parking lot.

By the time Morley got outside, they had the girl up against the squad car, arm pulled up behind her and twisted in the socket, screaming. Morley shouted at them to let her go. One of the officers barked at him to back off. Morley called him a choice name. The cop told him he was under arrest for disorderly conduct, grabbed something—Morley was never able to say afterward what that something was—and raised his arm above his head.

Morley ducked, and in one lunging, sweeping motion, upended him. The cop went down hard. Morley kicked him in the chest. The officer rolled over on his back, tried to cover his head. Morley reared back and kicked twice more, as though he were kicking a field goal, and the cop's nose and cheekbones caved like Chinese porcelain.

This time bail was set for twenty thousand dollars. Franklin and Carrs put up half each. Afterward, Carrs tried to talk Morley into fighting the charges. "I will go to court for you," the skipper told him. "That bastard egged you on. He was begging for trouble. That was police brutality." But Morley wouldn't have any of it. He didn't have the heart, not after the circus in Ketchikan. He pleaded guilty to third-degree assault and was sentenced to two and a half years, with the possibility of parole in fifteen months. He also had to pay the officer's medical bills and court costs—fifty thousand dollars in all—which left him in a twenty-thousand-dollar hole.

After his release, Morley got a job at a gas station a block from the Pioneer Bar, where he met Tamara Westcott. He moved in with her, but soon fell behind a few thousand dollars in his restitution payments. In the fall of 1997, his parole officer gave him an ultimatum: pay what he owed or he was going back to prison.

That was when Morley called Scott Echols and asked for a chance to skipper the *La Conte*.

Bob Doyle gazed out at the dark sea. "Well, I hope this all works out, Mark."

"It's got to."

"We'll do all right."

"Sure we will."

"Say, Mark—what are my chances of tendering with you in February?"

"I can't promise anything," Morley said heavily. "But I'll do what I can."

"I appreciate it."

They looked out the window a long minute. Then Bob Doyle said, "Why don't you get some sleep?"

"I'll try."

"I'll wake you if I need anything."

"Do that."

Bob Doyle stayed on watch until ten o'clock, and Morley was still awake when David Hanlon came up to relieve him.

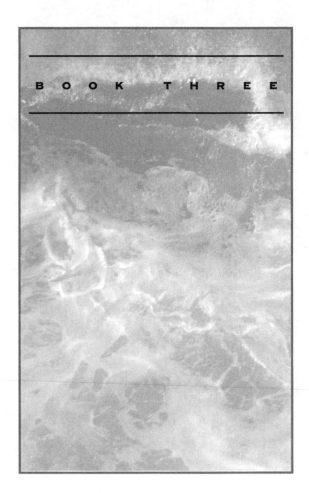

B O O K   T H R E E

By midnight they were approaching the grounds. The wind had freshened up, with the moon gone, and the current had swung around to the southwest. Since they were sailing against the tide the engine had to work harder to keep eight knots. The stars were gone now, and somehow the sky seemed lower, the low, dark sky of a squall coming. No other boats had shown on the radar for hours.

Once they made the Boot, a shoal that looked like a large shoe on the map, they set a course for the western fringe of the Fairweather Grounds, an area known as the Triple Forties. That was where a trio of seamounts rose from the depths of the gulf to within forty fathoms, or 240 feet, of the surface. Deep-sea currents upwelled at these mounts and brought up plankton, on which the tiniest of fish fed. In the evenings and mornings when there was a rising tide bigger fish would come into it to hunt, and in no time there would be halibut, black cod, quail back, tuna, lingcod, king salmon and yellow eye to choose from.

They were going to set over the shoals, starting on the lower bank. The idea was to release gear into the currents, from west to east, three strings at a time, and to drape the longline over the pinnacles. It was going to take six hours to lay out and haul back each set of three, two-mile-long strings. With luck, they would catch a thousand pounds of yellow eye each set. If all went spectacularly, they would pull twelve thousand pounds in four or, at the most, five sets, and be on their way to Juneau by suppertime on Sunday.

The marine forecast was calling for twenty-foot seas, gale-force winds and heavy sleet by early Sunday evening. They were not leaving much room for error.

The anchor made a big splash and main line zizzed through the chute. The line tightened as it hit the water and began racing out at a slant, slicing the water as it went.

"Anchor's out!" Gig Mork called up from the companionway. Mark Morley marked the drop position on his video plotter, leaned out the door of the wheelhouse and yelled, "Got it!"

The skate was running out in a blur, the gangions and hooks and line making that *woo-woo-woo-woo-woo* sound as they jumped through the metal chute. Main line was paying out into the ocean off to starboard. With the motor throttled up to six knots, the *La Conte* bumped and flumed through the chop. The line moved out steadily.

"One skate away!" Mork called up, and Morley gave him a thumbs-up sign. The others had bunched along the railing to watch the second skate as it rushed out. This one went into the water at a more urgent slant.

"Ease up on your drag a little!" Mork shouted to Morley. "Watch the line!" Then, to David Hanlon, "Get that third skate over here and get it tied on."

Morley throttled the motors down a touch. If he eased up just a hair on the gas the angle would lessen on the line. But he could not afford to slow the boat too much or main line would start looping and snarling. All this time, line kept shooting out taut and steady.

"Tight skate, this one," Mike DeCapua said.

"Real nice," Mork said.

"Who's is it?"

"Dave's."

"Nice fucking skate."

Bob Doyle looked off astern at the point where the ocean seemed to suck in the string. The line was racing into the curl of the waves that the wake raised. He could see the baits clinging tight to the hooks even after they exited the chute and he noticed the hooks glinting in the deck lights just before the froth swallowed them.

"Oh shit," DeCapua said, pointing. "This one's got to be Bob's."

The next skate was going *thunk-thunk-thunk-thunk* as it passed through the drum. Chunks of herring were whipping off the hooks and some of the hooks were embedded in the line itself.

"Sloppy."

"Most of the herring's off that one."

"That one ain't catching shit."

"Waste of good bait."

Even though he was standing in a raw, cutting wind, Bob Doyle could feel his cheeks and forehead burning in embarrassment. He had worked extra carefully and taken longer to finish his skates, hoping to impress the others. But most of the hooks on his skate were going into the water without baits. Hanlon tapped him on the shoulder.

"Where is your other skate?"

Bob Doyle nodded at the bait shed.

"Let's take a look at it."

The line kept jerking out and down, out and down. Mork and DeCapua were fastening a small anchor to the middle of the string to keep it on the seafloor, where bottom fish hid in crevices and caverns. In the meantime, Hanlon reworked Bob Doyle's second skate. He worked with the speed of a seamstress; there was no keeping up with him.

"Ah!" Bob Doyle yelped.

He had been trying to get the hook through the frozen salmon and the bait had slipped. The hook had gone instead through the flesh between his thumb and forefinger.

"Hold still," Hanlon said. He turned the hand over. "That's nothing." He tugged at the hook.

"Christ!"

"Imagine how it'd feel if your hands weren't cold. How's that?"

"Good Christ!"

"Watch yourself next time or you'll lose a thumb. You'll be okay. Go to the galley and wash that out good. Wrap it in gauze. Later on you can wash it out with alcohol or iodine. I'll handle the rest of this."

Just then they heard Mork call out. "You girls plan on getting me some skates anytime this year?"

The seas were eight feet and breaking a little higher now on the port bow and tightening the drag on the main line. When the line went too taut, Morley let up on the gas and coasted. When it went too slack he nudged the La Conte ahead as softly as he could so as not to catch or snap the line.

By the time Bob Doyle had come back out on deck, the fifth skate was almost done. Hanlon had brought out the rebaited skate, tied it off the last of the fifth and set it in a pile on the deck.

"Let's get a flag on the end of this one, here," Mork said to Hanlon. "Bob! Get me another pole marker near the bait shed."

"Right back."

Alongside the bait shed he found the aluminum pole marker. It had a flag on top that was easily picked up on the radar, and a float on the bottom. Their routine was to fasten markers to both ends of a string. That way they could track the drift of the string, or find both ends of the line if it parted. He carried the float up forward.

"Last skate in the chute!" Mork called up to Morley. "Last skate!" Turning back: "Quick, let's have that highflier."

His thick little hands worked fast, hooking the highflier to a hundred-yard stretch of line with a barrel knot before tying this line to the last of the outgoing skate. He stood up.

"Help me get this over," he said.

They hoisted the marker up over the gunwale and pitched it overboard. Then Mork and Bob Doyle tossed over the steel anchor. Mork called up, "Last anchor's out!"

Morley swung the boat around softly and easily so the stern hardly disturbed the calm sea. He stuck his head out of the wheelhouse.

"Good stuff!" he shouted. "All right, go take a piss or pull your puds or whatever else you do when you take a break. We haul back in twenty."

Mork and DeCapua went to the galley table for a smoke and a coffee, and Hanlon returned to the bait shed. Bob Doyle went to his cabin. He held his hand up in the light. It could have been worse. It could have gone through the palm. His fingers were cut and bruised and blistering. That was all right. He still had them. Not every deckhand could say that. He looked at the warm, sticky streak for a moment. Then he went to look for some Mercurochrome and gauze.

It was raining now and the first gusts of the squall were coming.

"This one's already fished what it's going to," Morley said. "Let's pull the fucker."

"All right," Mork said.

"Giggy," Morley said, "I want you to take the helm for this set, all right?"

"What for?"

"Nothing special. I just want to work that winch and the gaff a little. Why don't you take a break and steer a little?"

"Okay."

"And get Bob to pull that buoy out first."

"I'll get him on it."

Their first two pulls hadn't been bad. They had hauled about three hundred pounds of prime yellow eye. Some tigers and calico had come up and every fourth hook was a sand shark. There were also some halibut and lingcod. The lingcod most likely had spotted the smaller rockfish in distress on the hooks and attacked them. It was a real bastard of a fish. Even with the hook in its own mouth, the lingcod would still chew on the rockfish. Morley had to club them several times just to get them on board.

For supper they kept a few halibut, the badly mauled ones, but knocked off the rest. They were catching far more halibut than yellow eye, but by the opening rules, they were not allowed to keep any of it, only lingcod bycatch.

"Watch my hand signals," Morley said to Mork. "If you see me give you the slack sign, ease up on the throttle. And watch those following seas when you cut the engine. We're getting knocked from behind on some of them."

"All right."

"Also, let's not try to do this too fast. What you want is to keep the boat as parallel to the strings as you can. And keep an eye on the tides. They're running goddamned strong. Those buoys are drifting faster now to the west, too."

This third set had not been soaking even a half hour, but they had noticed that sand fleas and lingcod were getting at the catch so they had decided to begin hauling back sooner. They might not get as many bites as they would by letting the gear sit for an hour. But what they did catch would be worth more at market.

They went back to work, hauling the line out of the water at a thirty-degree angle. The first half-dozen hooks came up bare, and then they started to pull yellow eye. It was a pretty fish in the light of the search lamp, its gills as bright and orange as a hot poker, its body lithe and gleaming with white droplets, its eyes bulging and mouth working. The only sad thing was seeing the bladders; from having been surfaced so quickly, they had swollen like water balloons and popped and stuck out from their bodies. But that's the risk they took when they bit at the bait, Bob Doyle thought. They didn't know it then, but they sure know it now.

A large yellow eye, a thirty-pounder, hit with a *whump* on the deck. Morley whooped.

"Now, *that's* a fish," he shouted. "Damn, I'll bet we get forty bucks for her." He swung his arm at the wheelhouse so Mork would slow the spool.

"Giggy, back her down!"

Bob Doyle watched the skipper lift the bullhorn gaff with the oak dowel like it was a thirty-four-inch Louisville slugger, and then with a wide, flat swing, sink the curved metal hook into the yellow eye's head. It made a noise like hitting a pumpkin with a club. Morley jerked the fish off the hook, turned and cast it on the deck. It landed with a satisfying *whump*.

"Bob," he said, "all yours."

The yellow eye had only a shake or two left of fight. Bob Doyle stepped over, lifted the gills, popped free the serrated knife that dangled from a cord around his neck and plunged it into the fish until the blade disappeared. Before he pulled the knife out the eye of the fish had gone flat.

He kicked it down the hatch.

"Sweet Mary," Morley said, "yes, yes, yes. Here, we got another customer."

Now the real work began. All the miles, chores, discomfort, planning—all of it had been in preparation for this: haulback. They had to haul up two miles of line as quick as possible without mauling the catch. They worked furiously yet smoothly, timing one another as though they had been longlining together all their lives: Mork, keeping them at six knots and on a straight course; Morley, working the winch and snapping yellow eye and lingcod onto the deck; DeCapua, looping and coiling line, ripping off old bait and readying the hooks for more; Bob Doyle, killing, then sweeping fish into the hold and running the coiled line back to Hanlon in the bait shed so he could check it, splice it and rebait it. It took more than an hour to finish pulling the third string. No one was happy with it.

"Son of a whore," DeCapua grumbled under his breath. He slapped a yellow eye in a bin. He and Bob Doyle were icing the catch and lowering the totes into the forward hold.

"What is it now, Mike?"

"We didn't get a pound over a hundred on that pull," DeCapua said. "A fucking waste of time, that was. A fucking waste." He coughed and spat. "Where the hell's that guy got us setting, anyways?"

Bob Doyle did not answer him. "Hey, let's put some more ice on top of this row here."

"Ice we got," said DeCapua.

S I X T E E N

It was daylight now but raining harder and they could not see the buoy markers until they were nearly on top of them. They had laid another set just north of their first, had hauled in the first string and were motoring around to pick up the second. A heavy northeaster was blowing. Late in the morning the wind strengthened to thirty knots and began twanging the stay wires and shearing the tops of the waves. The seas had stacked up to eighteen feet and were breaking against the hull, throwing clouds of spray over the railings and flooding the foredeck silvery.

They were wet to the skin. White, salty rings crusted around their eyes and mouths. It was hard to breathe. The rain was sharp and driving gray and they could see only the roaring white of the breakers past the railings. It was as though a monsoon had opened up, or a sandstorm had erased the sun. The *La Conte* keeled sharply, shook with each big comber. When the deck pulled out from under them and dropped away they tucked in their chins, doubled over and leaned heavily into the roll to keep from skidding off. The squall was so violent and the rain so intense that trying to see the markers from twenty yards was like trying to see flags behind a waterfall.

The first string had gone badly. First a line anchor went missing, and then the wild, leaping seas began twisting the leaders and causing the main line to loop and come up in big knots. There was nothing to do but chop off and discard all of the snarled bits of line, pull off the hooks, tie and bait them again. They were going through an awful lot of bait. They were also losing large stretches of main line, though they did manage to salvage a good number of snaps and leaders.

"I swear to holy Christ," Mike DeCapua said, "this ain't no goddamned day to be making sets."

He had just walked into the bait shed and was shaking water out of his gloves. Gig Mork and David Hanlon were splicing and baiting gear. Bob Doyle was stripping ice off chum baits he had cut from a block.

"Bob," Mork said, "give me some more of them baits."

"It's crazy setting out in this crap," DeCapua said. "No, this ain't crazy. It's fucking insane." He looked at Hanlon and Bob Doyle. "What the hell are you guys doing, anyways?"

"What does it look like?"

"Baiting."

"Mike," Mork said. "Do me a favor and go down and check the bilges. They probably need pumping." DeCapua went out.

"I swear to God," Mork said to Bob Doyle, "one of these times I'm going to kill the fucker."

"He's not that bad."

"Oh no, he ain't that bad. He's fucking poison, is what he is. He's also a good coiler. If he wasn't, I'd already have thrown his candy ass overboard."

"Let him talk. He likes to talk."

"He talks a lot of shit."

"Let him."

"I just don't like him mouthing off to the skipper the way he does. Always asking questions. Always telling the skipper to do this or do that. Fuck that."

"He'll settle down."

Just then a huge swell passed under the boat. It lifted them up, up, up, the floor sliding out and away from under them and for a moment Bob Doyle felt out of himself, hanging in time and air, breathless, heart fluttering, dry-mouthed, like floating in a dark, cool, quiet water tank; and then the deck came up in a rush. The bucket he had been sitting on slammed into his rump as the boat walloped back into the ocean with a swell-splitting crash.

"Grab the lines!"

Seawater was flooding into the bait shed, sweeping gear and bait and coils of line back and forth.

"Okay, get them up here," Mork said. "Man, that was a nasty one. Dave, you all right?"

Hanlon looked at him and nodded. There was a paleness in his face that had not been there earlier. He was bailing water out the door.

"Christ," Mork said, "I can't stand this shit. All right, that's enough of that. Just—let's just hurry up and get these skates baited and tied."

DeCapua came in.

"Well?" Mork asked him.

"Water was up about a foot and a half. Don't worry. I pumped it all out. Wasn't that some swell? Monster son of a bitch. I'm telling you, these ain't no seas to be setting in."

"Anybody want a cigarette?" Mork asked.

"I'll take one," DeCapua said. Mork gave him an evil eye.

"Dave?"

Hanlon shook his head. Mork shoved the cigarettes in his shirt pocket. "Dave," he said, "I want you and Bob to finish this last skate. And you, Mike, get yourself ready to coil. We start pulling that second string in ten minutes. You hear me?"

He went topside.

They had the starter buoy aboard but the main line was coming in unevenly, in spurts. The seas were stacking higher still and the main line was going taut, then slack, then taut as the boat pitched up and down. There was a danger in having too much slack in the line; it was apt to wrap itself around bottom rocks and snap with the boat heaving as it was.

Gig Mork was on the winch, but he was having trouble finding a rhythm. He was cursing the fish and cursing the seas and cursing the machine. The first skate of the string was a near disaster. If they had pulled a dozen half-size yellow eye they would have been fortunate.

Mike DeCapua had just finished coiling the first skate and Bob Doyle was gathering the lines to run them to the bait shed when he heard the winch stop. Mork pushed past him.

"Hey, Gig." Mork did not answer him. He scuttled up the ladder to the pilothouse.

"Say," Bob Doyle said. "What's with him?"

DeCapua snickered. "Main line's fucked."

They still had seven skates out, all of them with new hooks and leaders. Mork came back down.

"So, what does the big man say?" DeCapua asked him.

"We're going to find the other marker buoy and pull the line in from that end."

"Okay," DeCapua said. "Now I got just one more question for you, Giggy."

"Shoot."

"Do we get overtime for rain?"

Darkness had settled over them by the time they spotted the floating flag of the end marker. It had drifted a quarter mile from where they had originally set out. Morley tried to steer them alongside the marker so they could fish the buoy out using the bull hook. But he could not get them close enough. The seas were over twenty feet by now, each swell bigger than the last, and the buoy kept dropping, then rising almost to within reach, then dropping away again. Sometimes they overshot the buoy and other times, when they threw out the engines to let the float catch up to them, they took combers over the stern.

At one point, everybody was out on deck throwing loops and line and hooks, trying to snag the float. Finally, Morley said, "Forget it."

"What?"

"I said to forget it. Let it go."

"Come on," Mork said, "we'll get it."

Morley shook his head.

"Let it go, Giggy."

Mork went topside to take the helm. The rest of them took a coffee break in the galley. They had a fifteen-skate string baited and had just started baiting another ten-skater.

Morley turned to DeCapua.

"All right," he said. "You two clear the deck and lash everything down. Then I want you back in the bait shed helping Hanlon ready that fifteen-skater. We're going to set it."

"*What?*"

"I want to put out another string."

"What for?"

"Don't ask me what for. We're going to get that fifteen-skater out and then we're going to head in to Graves Harbor. We already got half a string out there. We'll leave this one out there with it."

DeCapua scowled. "What's the sense of throwing out more gear if we ain't gonna pull it?"

"It's our last string. It's baited. It's going out."

"But we're going in."

"We're going to let these two strings fish for us and come out tomorrow night and pick them up. We got nothing to lose. The worst that can happen is we pull them up and get nothing."

"Oh, fuck."

"What'd you say?"

DeCapua had a cigarette in his mouth and was lighting it. He exhaled. "Nothing."

Morley gave him a steely look. "Anything else?" Bob Doyle shook his head. DeCapua glared at him.

"Get going, then."

Once Morley had climbed upstairs, DeCapua grabbed Bob Doyle by the shoulder.

"Hey, what's your problem?"

"How do you mean?"

"I want to know what your problem is. You're licking his ass so much I think you're starting to like the way it tastes."

"Go to hell, Mike."

"I ain't got far to go," DeCapua said.

The fourth string had been set and they were back in the bait shed stowing gear.

"You know," Mike DeCapua grumbled, "that's the last time we see those strings."

"What do you mean?" Mork asked.

"You know what I mean."

Bob Doyle said nothing. He was sitting on a crate. The nonstop rolling was working on him. His head was groggy and loose feeling.

DeCapua said, "Look at the seas out there, man. We're on the lower west bank of the Fairweather Grounds, setting straight out into the Pacific."

"So?"

"So the current is pulling hard. Real hard. That gear is gonna slide right off the edge of the shelf. There ain't no reason to come back and look for it. You know?"

"Sure, I know."

David Hanlon stood up and went for the door. "Be right back," he said, and ducked out.

"What's with him?"

"He's sick," Bob Doyle said. "I saw him lose it a couple of times already."

DeCapua turned back to Mork. "So the plan now is, we're going to Graves?"

"That's right," Mork said. "Graves."

"We should have been there already. We could be fixing this fucking boat's problems."

"Problems?"

"Like the stove."

"What wrong with the stove?"

"It don't work. Remember dinner?" Earlier, Morley had made them a halibut casserole with potatoes au gratin, carrots and peas, but it was cold.

"What about it?"

"Listen, I'm a cook. And that shit we ate wasn't cooked. That fucking thing hasn't worked right since it caught fire. And the generator is fucked. Which means we got no heat. Which means I'm going to be in a cold, wet bunk tonight. I'm wet all day, and then I gotta be wet at night."

Mork said nothing.

"Fucking boat," DeCapua said. "Leaks over my bunk. Leaks in the galley. Hell, I go to bed with the heater on and I wake up three hours later with nuts like ice cubes. Shit, not even the head works."

"Hey, that's enough."

"What?"

"You snivel too much."

"I got a degree in sniveling."

"Hey, listen," Mork said. "That's my buddy up there. He's a good skipper."

"Is that right?"

"That's right."

"Well, I'm having trouble believing that, Giggy. Okay? The guy could be a great cod fisherman. I don't know. We haven't tried to fish that. But I'm watching him rockfish and he's not doing the things that successful rockfishermen do."

"Like what?"

"Well," DeCapua said, "for one thing, he don't use rocks for anchors."

"A lot of guys use steel anchors."

"Fine. But he's always looking for pinnacles to set. That's what guys do when they first start rockfishing. You know? They always look for pinnacles, thinking, 'This is the *magic pinnacle* no one's ever seen before.'"

Mork said nothing.

"There is no fucking pinnacles out here with fish on them anymore. You know why? We *killed* all the fish. Killed them all twenty years ago when they paid us six cents a pound. That's right. That's why we had to fill our boats with fish to make a buck. Now they're two bucks a pound. And you know why? 'Cause there's none left."

He stopped.

"You done yet?" Mork asked.

"No."

"Listen, Mike," Mork said. "That last set—sure, maybe you're right. Maybe it's gonna drift off."

"See?"

"That ain't the point. What matters is we only got a few more days to fish. And then we got to take the boat back to Juneau. So let's just fish. Okay? Let's just fish."

DeCapua said nothing.

"Get some sleep," Mork said. "I got first watch. Then you, then Bob, then Dave."

"Just one more thing," DeCapua said.

"No," Mork said. "No more. Finish. *Finito.* The end."

"All right. But later—"

"Okay," Mork said. "Sure. Later. But not now. No more from you right now."

SEVENTEEN

They drifted offshore a half mile in the dark. The swells had calmed down a little and the tide was running in fast. They could see tiny keys at the mouth of the bay, rising out of the water like black hedges, and hear the faint crashing of surf on the shore. The wind was still twanging the wires and the boat lurched with each slap of wind.

"I can't figure it," Gig Mork said to Mark Morley. "You better go down and take a look."

"Fuck."

"Keep your head on. It's gusting a little and we got no steerage right now. But these seas ain't that big."

"What do you think it is?"

"I don't know. We got plenty of gas."

"And the generator?"

"It quit with the engine," Mork said.

They had no electronics or lights so Morley studied the charts with a flashlight. Judging by the chart, they were drifting in fifty fathoms of water.

"What time is it?"

"Almost three-thirty in the A.M."

Graves Harbor was the first good cove to lay up in along the Fairweather coast south of Yakutat. It had two arms. The South Arm was wider and better protected. The North Arm was more exposed to the gulf. But both channels had plenty of rocky heads. Around the bay were spruce and pines and above them the Fairweather Range, standing so high and bulky and white that sailors got dizzy looking at it from up close on a clear day.

It was a pretty bay to ride out a storm in the daytime, but it was no place to be at night without a generator to run radar, a searchlight or a depth finder.

"Watch her," Morley said. "I'm going down for a look. When I'm ready I'll give you a signal to start her."

He went down below.

"Bob," Mork said.

"What?"

"I want you guys to keep an eye out for shoals or rocks or any other nasty shit."

"Okay."

"You see anything, you holler. You got that?"

A few minutes later Mike DeCapua came up the outside ladder and poked his head in the side door. "Skipper says for you to roll the engine over, then to wait ten seconds, and then to do it again."

"Okay."

Mork hit the starter key. The engine chuffed and coughed but the motor didn't roll over. He tried it again and again but it would not catch. Again DeCapua came up the ladder.

"The skipper says to hold off." They heard boots along the companionway, and then Morley shouted to them from below on the foredeck. "That lousy piece of shit motor ain't getting any gas," he said. "How's our heading?"

They had already passed the island at the mouth of the cove and the swells were brushing them in toward shore.

DeCapua said, "There's rocks all over the place in here. This is no place to be without a sounder."

Mork put his head outside and shouted, "I say we get the anchor down right away. We got to stop our drift."

"Then get on out here and get the anchor down," Morley called to him. "There's forty, maybe fifty fathoms on that anchor line. At some point it's got to catch. We're in trouble if we get anywhere near that beach."

"Coming down."

Bob Doyle said to Morley, "Hey, listen. I got an idea."

"What is it?"

"How about getting some tarps up and making a sail in our forward section?"

"How's that?"

"We can rig some tarps up between our cranes and booms up forward, real high up. They'd make a decent sail. It's gusting all around but there is a steady breeze blowing offshore. The tarps might catch that wind and keep us off the rocks."

"That's a fine idea, Bob. Okay, go get Mike and Hanlon and have them help you. Where the fuck is Hanlon anyway?"

"In the bait shed."

"Well, Jesus Christ, tell him to get the fuck out here. We need him. If any of you guys so much as sees a pebble sticking out of the water, I better hear yelling."

Bob Doyle nodded.

"And one more thing," Morley said. "What do you think as far as the Coast Guard goes?"

"What do you mean?"

"I mean, we may have to call them at some point. We're not a navigational hazard or anything, but we might blow up on the rocks."

"I'd call right now. Let them know to stand by."

"Think they'll come out in this shit if we need them?"

"Sure," Bob Doyle said, "they always got a ready crew."

Morley looked at him. "Do me a favor, Bob. Get them on Channel Sixteen. Give them our position. And tell them we may need them soon."

Morley was in the engine room and the others were hurrying to rig up the tarps. They were having a time of it. The boat was still pitching badly enough. Not so distant now, they could hear waves beating on rocks.

Mork said, "How far are we?"

"Half a mile," DeCapua said, "no more."

They punched holes in the tarps, looped rope through the holes and secured the tarps to the booms and crane. Once they hoisted the tarps the wind filled them and the boat leaned to starboard.

"It's working! It's working!"

"Shut up!"

It was hard to tell, for sure, how fast the boat was drifting; they could not even see mountains. Bob Doyle scampered up to the wheelhouse, updated the Coast Guard and then rushed down to the engine room to find Morley hunched over the engine block with tools and rags and tubes scattered all around.

"How's it coming?"

"Almost there."

He climbed up on deck.

"How's he doing?" Mork asked him.

"He thinks all the knocking around loosened up sediment and shit in the fuel tanks. It probably clogged a line or a filter."

"How close is he?"

"Close."

There was nothing to do but stare into the blackness and listen to surf smacking on the rocks.

Mork called out to DeCapua, "How far?"

"Anytime."

The tarps flapped and whipped and the boat was leaning heavily now, and Bob Doyle was thinking that any minute he would have to make a dash for his survival suit when suddenly he heard a deep gurgling sound and then a grunt, and then a whoop, and finally the sound of pistons firing.

The skipper woke them a little before eleven. The strong southwest wind that had come up before dawn had weakened to less than a gale. The boat was rocking less now and a big surf was piling on the beach.

Breakfast was pancakes, browned, with an egg on top, four strips of bacon, buttered toast, bagels with peanut butter, coffee and milk and a big cup of chilled grapefruit juice. Everyone ate heartily. They were hungry and the wind gave them even more appetite.

"Who wants more coffee?" Morley asked. He had just finished serving the last of the bacon and toast.

"Right here," Bob Doyle said.

Morley refilled his mug.

"I want to check the fuel lines again," he said. "And then after that I want to try and see if we can pick up that gear we left out on the grounds. Can't leave them out there too long. Sand fleas will eat everything we caught."

No one said anything.

"By the way, I called the owner, told him where we were and let him know how we're doing."

"When was this?" DeCapua asked.

"This morning," Morley said, "while you were sleeping."

After breakfast Morley ran another check of the fuel lines and filters and took apart the main generator. He found a few loose, wet parts. While he did this the others worked on mending gear. Sometimes Mork went below to give Morley a hand. Bob Doyle chopped bait loose from the ice. Hanlon spliced line. DeCapua worked on some hooks and showed them how to roll cigarettes one-handed, and he whistled to a few songs on the stereo and even told a few jokes.

Sleep had smoothed over the tension and bad feelings from the previous night. Mork and DeCapua acted as though nothing had happened. Hanlon was reserved, his expression a bit taut and sallow at times, but he cheered up finally. Mork was good at telling sea stories. So was DeCapua. He was witty. It was almost as though they were a bunch of guys on a vacation from their wives the way they got on so well together.

E I G H T E E N

They pulled anchor just after two that afternoon. Circling gulls watched them coil the dripping rope. The sky hung like bunting over the mountains and there was a mist in the pines behind the beach. The marine radio was still reporting fifteen-foot seas out in the gulf. The storm was moving slowly north and east, the forecaster said, and there was a warning to boats to watch for gales. They laughed when they heard that.

The tide was going out, and Bob Doyle stood near the railing and looked down at the water, as gray as the sky, and listened to the bow part the chop. As soon as they slid out of the bay he heard the deeper hum of the engine, felt the bow lifting and the wind coming sharper and colder and, squinting into the wet wind, saw the horizon dark and not ugly. He blew on his hands and went to the galley for a coffee.

They dogged down the hatches, stowed main line, topped off the gas-powered bilge pump and the main engine and lashed the fuel and water jugs to the whaleback. Bob Doyle was pulling out the scupper plates when he heard a loud hoot and laughter coming from the wheelhouse.

A few minutes later Gig Mork entered the bait shed. Bob Doyle, hanging coils of line, said to him, "What's all the noise about?"

Mork lit a cigarette.

"Ah, just the skipper," he said. He inhaled comfortably and let the smoke slip from his lips. "He got a call from his girlfriend."

"Good news?"

"You could say."

"Well," Bob Doyle said, "what was it?"

"She's gonna have a baby."

"Really?"

"That's what the man says."

"Wow," Bob Doyle said. "That is great. Jesus. When's the baby due?"

"I don't know. Sometime this year."

"I'll bet she's happy."

"Well," Mork said, "she's had a time of it. See, the doctor first told her that she had lost a baby. And she did. Only it turns out she was gonna have *twins*. Just didn't know it."

He puffed on his cigarette.

"No shit," Bob Doyle said. "Talk about second chances. How did Mark take the news?"

"He's one happy son of a bitch. I can tell you that."

"Well," Bob Doyle said, "that's some news."

Right after they passed the fifty-fathom curve the seas stacked, and by four that afternoon curlers were pounding the stern and washing over the decks. The wind howled like a gut-shot gray wolf and the boat was keeling almost forty degrees on the bigger swells.

Once Bob Doyle finished stowing the gear he staggered dopily to his room. He did not have his sea legs yet. His thigh muscles fluttered and his ankles were sore. He wondered if he ever would get his sea legs. He eased himself down on his bunk. The porthole looked like the door of a washing machine. He thought he would time the waves. He had counted to eight when the door swung open and he saw Mike DeCapua, holding a five-gallon bucket with water in it.

"I gotta take a crap."

"Lovely."

DeCapua pulled down his sweatpants, yanked off his rubber boots, sat down on the bucket and thumbed to a page of a paperback he had dog-eared.

"Mike?"

DeCapua looked up.

"Can you put your socks on or something?"

"Why?"

"Your feet stink. I mean, they really, really stink. They're fucking horrible. You ever think about scrubbing them with lye or something?"

"Shut up."

"Your shit's gotta smell better."

"Fuck off."

Bob Doyle closed his eyes. It felt as though his bunk was shoving him back and forth. After a minute he opened his eyes and said, "How long is this gonna be?"

From behind his book, DeCapua said, "I don't know. Could be days. You got any cigarettes on you?"

"No way."

DeCapua lowered the book. "Okay," he said, "but I always crap faster with a smoke."

Bob Doyle threw a pack of Luckies at him. DeCapua picked the pack up, drew out a cigarette, lit it and went on reading.

Later they were up in the wheelhouse. The sky looked silvery now in the afterlight and was letting go in black streaks across the gray horizon.

Mark Morley said, "We've been taking a lot of waves on our stern, haven't we?" He paused. "How are the bilges?"

"We've been pumping them every hour or so."

Morley frowned.

"These damned seas aren't getting any better," he said. "We still got another forty miles or so." Bob Doyle noticed the dark pouches under his eyes twitch.

Mork said, "So what do we do?"

"Turn around," Morley said. His voice was toneless, flat as a piece of slate. "It ain't worth pounding through this shit. Not for a couple of lines."

DeCapua smiled a faint, economical smile.

"It's too bad," Morley said, "but I don't see any way around it. Shit." He sighed.

"Giggy," he said, "turn her around."

"All right."

"I guess we'll go back to Graves," Morley said.

"Skipper," DeCapua said, "why do we have to go to Graves? I mean, there ain't nothing there. It don't make sense to sit in a bay for two days."

"What's your idea?"

"I say we go to Elfin Cove. It's another fifty miles, but it's a neat little town. And they got parts and fuel there. And fresh water, too." He paused, then added, "And, there's always a chance we can scarf some beer or hair pie."

"I could use some of that," Mork said.

Morley thought it over.

"Giggy," he said, "set a course to Elfin Cove."

DeCapua clapped.

"Say, there, Giggy," he said, "you been to Elfin Cove plenty of times. What kind of shit they got going this time of year?"

"You guys go talk somewhere else," Morley said. "Better yet, get some sleep. And don't forget. I want those bilges checked every hour."

Bob Doyle and DeCapua went below. They did not find Hanlon in the galley.

"Finally some reason in a world of insanity," DeCapua said. "Hey, where the hell's old Davie boy?"

"In his rack."

"Just wait till he hears the good news."

N I N E T E E N

They limped into Elfin Cove in the dark and drifted to a stop alongside a fuel dock. The air hung cold and still, and in the channel the water lay flat and black. All around rose mountains, steep, muscular mountains, their tops jagged with spruce and dusted with snow. There was a path of wood planks that began at the far end of the dock and twisted up a slope to the top of a ridge. Behind the ridge was the soft, yellow glow of town lights, and they could see ropes of smoke twisting high above the trees. Nothing stirred otherwise. It had snowed and everything had a stiff, frosted look.

Bob Doyle stood on the foredeck and took in the mountains. Spruce look so lifeless with snow on them, he thought. Like upside-down icicles. Or glazed pipe cleaners. Sure snowed a lot. He swung his legs over the gunwale and dropped down on the dock. His legs wobbled a bit. The motion of the big seas was still in them.

Mark Morley came out on the foredeck.

"Quiet," he said.

"Too quiet."

Morley nodded. "Okay, let's get the anchor down and tie ourselves up."

They tied off the lines to the bow and stern, and David Hanlon wedged a few buoys between the hull and the pier. Gig Mork dumped two anchors. The anchors went in with a small splash and sent ripples across the channel. Morley came up the fo'c'sle where he had gone to stow the rope and tarps.

"Looks good," he said. "All right. Why don't you fellas take the night off? Only don't go getting in any trouble. I'll see you in the morning."

Mike DeCapua came out on deck. He had just woken up. His face looked like a mass of gray putty with black stubble on the lower part of it.

"Elfin Cove?"

"Yup."

"You going into town?"

"I guess so."

"Hold on a second, will you? I gotta get something." He went back inside and came out a few minutes later with a bag under his arm.

"What's that?"

"My doobie."

They walked along the slippery, planked path and came to a framed cabin with a wood sign: COHO'S BAR AND GRILL. Another sign hung on the front door.

NO RAINGEAR NO MUDDY BOOTS PLEASE

There were square tables and chairs with shiny tops and a bar along the wall with gleaming liquor bottles and a stainless-steel grill in the far corner. Very tidy. But there was no one inside.

"Time's it?"

"Almost eight."

DeCapua snorted. "Who pays to drink, anyway?"

Elfin Cove was a nice-looking little town. Wood-and-stone cabins with chimneys clung to the hillsides, and there was a cove with a quay and little boats in the slips and nets hanging from racks. A few cruisers had their cabin lights on. They glowed like candlelit pumpkins.

"Gonna call your kids?"

"Right now," Bob Doyle said.

They walked along a boardwalk, past a post office and what looked like a little school and stopped on a ramp in front of a blue phone box.

"It's the only public phone in town. Everyone uses it. You can make your call from here," DeCapua said. He pointed to a garage that had a white light around the door edges. "That's The Shop. I'll be in there. Come on in when you're done and have a smoke."

"Okay."

Bob Doyle lifted the receiver. The number he dialed rang four times.

"Yeah?"

"Is Laurie there?"

"No."

"Where is she?"

"Out."

You bastard, Bob Doyle thought. You insolent bastard. "If the kids are around," he said, "I'd like to talk with them."

There was a grunt, some muffled sounds and then shoes tapping on a floor. Then he heard the voice that sent a jolt through him.

"*Daddy!*"

"Hello, sweetheart."

"What are you *doing*, Daddy?"

"You know what I'm doing, silly. I'm fishing."

"Oh, sure." There was a pause. "Daddy, tell me a Barbie story."

Whenever he was away, Katie expected him to call in every day with

the latest episode of Barbie's Alaska adventure. On their way to Elfin Cove
he had thought up a new one: Barbie Fishes the High Seas.

When he finished, she giggled. "That was *good.* Tell me another one."

"No, honey, that's all for tonight. Is your brother there, sweetie?" He
heard more muffled sounds, some yelling, and then crying, and then his
son's voice.

"Dad?"

"Brennie, did you hit your sister?"

"No."

"Try to be nice to her."

"How are you, Dad?"

"I had a great day fishing. We caught lots of stuff." He paused. "What's
going on in school?"

"Not much."

He was not very good at chitchat so he told his son about the trip. He
told him how big the yellow eye were, how big the waves had been, told
him names of different rockfish, skipped over the size of the catch. Had he
seen any whales? Not yet. How about sharks? None of those either. Would
he bring back some shark teeth? With any luck, yes. When was he coming
home?

"As soon as we fill our holds."

"And then?"

"Then I'll fly down to Sitka."

"What if you can't?"

"Then you and your sister will have to fly up to Juneau to see me."

"Cool!"

"Brennie," Bob Doyle said. "I gotta go now."

"Will you call tomorrow?"

"Sure. How about eight o'clock?" He knew the children usually were
in bed by nine.

"I love you, Dad. Hurry home."

Bob Doyle hung up. The sound of his kids' voices used to fill his
insides, at least for a little while. Not anymore. Now that he knew he had
no home to return to, it only made him feel hollow and lonely.

It was cold in the night and Bob Doyle slept heavily. Once he woke and, moving his shoulder against the hard bunk, realized he was back on the boat. He pushed his legs out as far as they would go in the sleeping bag, and stretching, the calve muscles hard as stones, the lower back muscles stiff and sharply aching and smelling the old herring and seawater in the crackling hair of his beard, he felt the pleasant nip of dry, cold air on his nose and the draining, splendid emptiness of fatigue.

It would be nice to do a hot tub in Tenakee Springs right now, he thought. But I had many chances to soak my body before starting this job and I am really fishing now. There will be plenty of chances to go tubbing when we return the boat. And I haven't missed the drinking yet. That's good.

He pulled the edge of the sleeping bag up to his nose, and lying in the dark, the warmth coming back to his split, blistered lips, he tried to let himself slip down steeply into sleep.

Now in his mind he saw the two-story house on Davis Street with its gutters piled with snow and the yellow lamplight coming through the front bay window. It was early for such a snowfall in Vermont and it had caught the Bellows Falls Fire Department by surprise and the trucks hadn't been out to clear the streets. Everything was so white and clean-linen looking. It was still falling, lazily, the flakes dropping diagonally through the bare branches of the oak and melting on his cheeks and forehead. He was standing out in the front yard, just beyond the reach of the porch light. He couldn't quite see his mother's face. Her arms were crossed.

Where have you been?

In the canal.

The canal?

He only nodded.

And that?

He held a tiny cat, pressing it tight against him. The cat was soaked. Its black hair clung to its bony back.

Do I get an answer?

It was in the canal.

In the canal?

Yes.

And you went in after it?

Yes.

Oh, God. Don't tell me you went into the canal near the power dam. You did, didn't you? Oh, God. Oh, oh, oh.

She put her hands to her face and turned and went inside. He stroked the wet cat. After a few minutes his mother returned with a big white towel. The sternness was gone from her voice.

Come here, Bobby.

He did not look up as she wiped the cat with the towel.

Weren't there any grown-ups that could have jumped into the canal?

Yeah.

Go upstairs. I've started a bath. Get out of those wet clothes and get in that tub.

He hesitated. His mother looked at the cat. It was a sickly looking little thing. Not a fluffy, fat cat.

All right. She sighed. The cat, too.

At first light he woke again and after breakfast walked into town to fetch drinking water. They spent most of the morning hauling buckets of water from the community hall to the boat. Once the freshwater chest had been topped off, Gig Mork spread the oldest, rattiest gear they had out on the dock and told them to pull up a bucket and get to work repairing it.

Mark Morley had gone off to pick up some extra engine parts and to buy some new tools. He had flown into a rage that morning while fixing the stove. He could not figure out why the burners would not ignite. After dismantling and reassembling the thing several times, he stood up and began hurling tools into the channel. He threw wrenches and adjustable sockets and screwdrivers and clamps and pliers and he had grabbed the stove itself and was hauling it out on deck to toss it overboard, too, when Mork ran over and stopped him.

By the time the skipper had gone into town the sky had turned to a dull, uniform gray so that the sky hung soft and heavy, cutting off the tops of the mountains. Soon a light snow was falling. The flakes dropped in swirls, not diagonally like hail but softly, like rose petals, and the crew sat on the dock, untangling and splicing line and watching them come, circling, from above.

"Why doesn't it snow out at sea?" Bob Doyle asked.

"It does," Mike DeCapua said, "only there's too much humidity near the water. By the time the shit gets down low, it's already turning to rain."

"Too bad it doesn't snow out there like this. It's pretty."

David Hanlon stood up, climbed onto the boat and went back to the bait shed. "He never talks much, does he?" Bob Doyle said to Mork, who was chopping a snarl off the line.

"No."

"He's got a funny walk, too. I just noticed it."

"His back is fucked. That's why he don't winch or gaff. He's strong as shit but he's got that back."

"How did he do it?"

"Says he came in from a trip once, in Ketchikan, and it was dark and he climbed up the ladder and didn't see the hole in the pier. In went the leg, all the way up to his balls. He never got over it."

"Nasty."

"Takes all kinds of pills, gets all kinds of treatment, but you know doctors. Take your money, never make you right."

"Man."

"Hand me one of those hooks."

A troller was moving up the channel. The men on it were sweeping the snow off the railings. They waved at them.

"Bob," Mork said, "don't make your splices so fat. They can't be that big. Try to keep them smaller than the diameter of the line."

"Sorry."

"And that section there is fucked. We can't use it and we ain't going to make it right. Here. Use this and cut it off."

They kept working as the flakes circled and drifted down around them. They smoked and told stories to pass the time and at one point Mork asked Bob Doyle if he had ever been to Pelican and Doyle told him he had passed by it a few times but had never stopped there. Mork said it was as good a town as any you could find in southeastern Alaska, that when you lived there you felt a part of one, big family. Nobody was ever in such a hurry not to stop and say hello and everybody who was part of the family got a nickname at one time or another. Bob Doyle asked him how they got a nickname like Gig out of a first name like William.

"First word that come out of my mouth," Mork said. "*Gig*. My parents just started calling me it."

Pelican, he said, got its name from a packer boat. That was a century earlier. His grandfather was one of Pelican's first settlers. His name was Nels Hjlmer, and he was a Norwegian who came to America on a merchant's ship at fifteen. The army officials could not pronounce his last name, so he gave them the name of his family's farm in Norway, Mork. Later, he took a Tlingit woman for his wife, had seven kids and built a big farmhouse for them. He never finished the inside. Too much fishing, too much drinking. After old Nels split with his wife, she and the older kids went to work at the packinghouse, sliming and freezing fish. Twelve-hour days. For a summer's work, the kids got paid twelve dollars and a pair of boots.

In those days, he said, the fishing was good everywhere. The seiners jammed the docks and the fishermen packed the cathouses. There was a big float house moored just off the pier that featured lots of kittens. After selling their catch the deckhands always rowed straight for it. One fisherman, the story went, fell hard for one of the kittens. One night he returned from a trip and found her servicing another client. No, he did not do anything stupid or make a mess; he untied the float house from the pier, towed it into the gulf and let it go. Just let it go. The float house drifted for two days.

Finally, some of the other fishermen starting feeling bad for the kittens—and lonesome—so they sailed out, hooked up the float house and towed it back to Pelican.

"They must have been happy to see those fishermen," Bob Doyle said.

"Nobody got a freebie, if that's what you mean," Mork said. They all laughed. Hanlon returned with a bucket of hooks and sat down.

"You ever been to Rosie's Bar and Grill?"

"Not yet," Bob Doyle said.

At one time, Rosie's was just a warehouse. Then somebody turned it into a bar. Then Rosie bought it. She decorated it real sweet, with fishing gear and antiques and big, old wood tables and iron lamps and a huge sign outside with the name carved into it by a few loggers. It was big and dark inside and smelled of stale beer and sawdust. There were bunks in a cabin out the back and Rosie let the miners crash there if they had nowhere else to stay. On real busy nights, she had guys sleeping on the pool tables and in the booths.

It was more than just a bar, he said. It was a radio hub for fishing vessels. A marine radio was set up at the bar so that skippers could call ahead to place a supper order, or to let Rosie know about an emergency, or a serious injury on a vessel. It was the place fishermen without families passed holidays. It was where they held their funerals, baptisms, weddings. Rosie's youngest, Sassy, tied the knot right under the brass bell.

When Gig Mork was old enough, he tended bar at Rosie's in the evenings. He wore a white shirt with pearly buttons and blue jeans, a red bow tie, red garters and a white apron. If there were more than five ladies in the bar Rosie would politely ask him if he would not mind giving the girls a little show, and he would have a few drinks, then a few more, to help him over his shyness, and before long he would be up on that bar flipping off one shoe, then the other, then one sock, then the other, and, gyrating his pelvis, peel off his shirt and jeans and let them fly.

"You're kidding me," Bob Doyle said.

"Hell, no."

When he was in a groove, he would swing his hips and bump and grind and played peekaboo in such a way that a few of the really juiced ladies tried to pinch him. One time while he was dancing a woman snuck up behind him with a pair of scissors. Rosie grabbed the woman and restrained her just as she was about to pounce; apparently, the customer wanted to snip the strings holding up Mork's apron.

"Hell," DeCapua said, "why do women always screw up a good thing?"

Then DeCapua told them about some of his most colorful capers, including the time when he was fifteen and he hit Andy's Foodtown in Hartford. His plan was to hide in the crawl space above the ceiling until the supermarket closed, then slither along the ceiling to the manager's office and drop down on the safe—so as not to trigger the floor alarm.

"How'd you get the combination?"

"I watched him open the safe from the crawl space," DeCapua told them. "I had binoculars."

"How'd you get caught?"

Waiting for the employees to clear out for the night, he did a whack of heroin and, as he was nodding, leaned back and tumbled through the ceiling tiles. He landed smack in cold cuts.

"Worst part was I hate fucking baloney," DeCapua said. Everyone laughed.

Then Mork told them a story about the night he drank himself to sleep while bartending at Rosie's, only to wake up in a booth wearing nothing but a sheer, red baby-doll nightgown. At that, Bob Doyle commented that there was nothing quite like the feel of satin once you were shit-faced. DeCapua asked him how he could know of such things, being as he could not get his own wife to put out, to which Bob Doyle replied that at least he did not chase hookers away simply by removing his boots. DeCapua denied that was true, but added that if a whore had ever complained he would have told her, "You're just smelling money, honey."

"Money?" Bob Doyle said. "To me you smell like a monkey with no money."

"Listen, shiny head," DeCapua said, flapping his ponytail at him, "at least something's growing on top of my head."

"Fuck your ponytail."

"Baldy here is mad," DeCapua said to the others, "because cue balls don't get pussy."

"And you do?"

"Ah," DeCapua said, "who the hell wants pussy, anyway? Pussy. Let me tell you something about pussy. It's totally overrated."

"What?"

"You heard me. Overrated. Who in hell needs to fuck? All that does is get you child-support bills. I'll attest to that. Kissing, too. It's all shit. Love. You know, World War Two was when they started in with all that love crap. Prior to that, in the movies, if you married a guy you were stuck with a guy. The good guy in the flick had to kill the bad guy to get the wife from him. You know? But then after World War Two they started telling everyone in America, 'You gotta be in love. You shouldn't stay married to him if you

don't love him.' So now they get married for *love*. And then what happens
in a month? They realize they ain't gonna change the beer-drinking son of
a bitch. Then they say: 'I don't *love* him anymore. So I'm going to move
out and take the kids.' And the kids suffer. We got this idea that *I must be
in love all the time*. Well, sorry, baby. I don't care how much you think you
love him. There's going to be mornings you're going to wake up and not
even *like* him. And nowadays when that happens, what do women do?
They get a divorce."

He had a captive audience now and he knew it.

"Marriages overseas are still arranged," he said. "Both parents have to
give a blessing or it's over. Here, we got Jerry Springer and Oprah telling
everybody how to stay married. What a joke. Now, let's say you go to have
sex with your wife, and she acts with reluctance. Okay? Well, if you're a
man, the way you handle that in Europe is, you quietly go off and get somebody on the side. You work a few more hours in the office for the money to
pay for a mistress. And you don't bring it home and you don't rub it in her
face. And your old lady goes out and gets a little on her own, and she don't
bring it home and make an issue of it. See? There's ways to get around the
love thing. Ways that have been working for hundreds of years."

"You done?"

"None of this *honey* this or *honey* that for sex. Fuck that. I'd rather pay
for it and get what I want." He paused for a breath. "And quit giving me
shit about my feet."

"They could use some soap and water."

"Fuck *that*. The only thing I need cleaned is my knob. And bitches are
going to take care of that, anyway. They ain't gonna be sucking on my toes.
Which leads me to my final point. You know what the two best things in
life are?"

"No."

"Blow jobs and beer—in that order."

"Oh hell."

DeCapua stood up.

"Where you going?" Mork asked him.

"I'm sick of this shit. I'm going into town. This is busywork. I ain't here
to do busywork. We got plenty of gear. I'm a fisherman. I'm here to fish."

Mork glared at him. "Sit down."

"I'll work when I'm on the boat but I ain't on the boat right now."

"You're on the clock. Sit down."

"This is make-work. Make-work for the crew. Hell. I just quit Phil Wiley. This is Phil crap all over again. Except when I was with *that* Phil, I made money. This is Phil Junior, and he don't make money."

"We're all a part of this," Mork said. He wagged a hook at him. "Let's get something straight here. On this boat, if the skipper tells you to take a shit off the stern, then you drop your stinking sweats and do it. You got it?"

DeCapua's face went as blank as a pie pan. "So what do I do if I ain't had a meal?"

He laughed.

"You know something, Mike?" Mork said. "If we do make money, I'm going to fine you."

"Oh, right."

"I'll take it out of your share."

"If we ever get one."

Just then, Morley came up the dock. He was carrying tubes of grease, tools and fittings.

"Hey, fellas," he said. Nobody looked up. "How's it going out here?"

T W E N T Y - O N E

The next day the sun broke through the clouds and there were no more rows. Everyone wanted to get the boat ready for fishing. Gig Mork borrowed some lube oil from a friend, Jim Lewis, who ran a parts store in town, and spent the rest of his morning tinkering with the generator. Mark Morley lubed the driveshaft, bled the fuel lines again and timed the engine. Around lunchtime the two of them looked for a backup generator to rent, but they had no money for a deposit. They also asked around for a six-man life raft. They were supposed to have one on board, according to

the law, if they fished open waters. But there were no big trollers in dock
and nobody along the quay had such a raft.

On the dock David Hanlon taught Bob Doyle how to better splice and
add hooks to lines. They fixed up some of the old, tangled gear and thawed
out more bait. Then Hanlon showed him how to salt down the chums in
five-gallon buckets so they would not spoil. They repositioned the fifty-five-
gallon fuel drum in the engine room, organized the galley and lashed and
stowed all the gas jugs, tools, longline, buoys and bait crates. Mike DeCapua did not get up until ten that morning. But later on he chipped in, too,
splicing line by himself on the foredeck. He even patched together
enough of the frayed gear to build a string.

They also inspected the boat for leaks and hazards. In the lazarette they
noted some seepage. But the leak, Morley decided, was not anything their
bilge pumps could not handle. They moved the gas-powered bilge pump
downstairs and bolted it to the floor beneath the ladder in case they started
taking on a lot of water. There was a loose plank on the hull near the prop
that had been lead-patched, but the patch did not look in great shape.
Morley told them that if it started leaking badly, the bilge pumps would be
able to handle it just fine.

The fo'c'sle was a mess. There were mooring lines, buckets, tarps, pallets and other odds and ends scattered about, and when they reorganized
things they noticed that planking had buckled on the starboard side. There
was water on the tarps. It might have come through surface cracks on the
deck, though DeCapua pointed out that the hatches were not watertight
either. No matter how they lashed it, he told the skipper, they were going
to take on some water if any waves came over the bow. It was not an ideal
situation, having leaking hatches. But nobody spoke up when Morley told
them to towel the fo'c'sle down.

During the afternoon Mork went off into the mountains with Morley's
hunting rifle. The skipper stayed on the boat and tinkered with the stove
and talked on the phone. Bob Doyle walked around the town and met a
few people along Potter's Quay. He wound up at the community hall
where there was fresh coffee on. He smoked and read the Juneau papers.
Hanlon called his brother in Hoonah, spoke with his nephew, Jimmy, and
caught up on some family news. DeCapua visited a friend on his troller,
and later visited his buddies at The Shop. It had a TV and a scanner that

allowed you to listen to radio conversation between boats at sea. DeCapua listened to see if any longliners that regularly fished the Fairweather Grounds were headed out there. None were.

It was a nice day, sunny, in the low fifties, with high, fleecy clouds. By the time Mork returned from hunting, the clouds were streaked in hues of gold and pink and all of the mechanical repairs were done, the fuel topped off. Mork told them he had not even gotten a shot off.

The chill of the night returned and everyone was looking forward to a big hot meal and a good night's sleep. It was hard to be upset about anything on a day like that.

Supper was excellent. The pot roast was browned and smothered in gravy outside, tender and juicy on the inside. They all sopped up the fatty juices and gravy from their plates with buttered biscuits. There was potato salad with lots of chopped-up boiled egg, mayonnaise and rough-ground black pepper in it, and they had carrots and cabbage, steamed and sliced, then cooked in butter and smothered in fried onions. Mark Morley and Gig Mork drank bottles of Coke. David Hanlon and Bob Doyle had milk. Mike DeCapua hated milk and did not care much for soda; he stuck with black coffee.

Everyone was hungry with the clear, chilly air and they all took second helpings. They were so hungry that nobody said anything when DeCapua remarked that it was foolish to return to the Triple Forties to pick up the gear.

"Anyone want more carrots?" Morley asked.

"Right here," Bob Doyle said.

"I can heat up some of that leftover spaghetti and marinara sauce. What do you say, Bob?"

"Not for me."

"Gig?"

"No, thanks."

"Dave, here's your roast."

By the end of the meal there was very little pot roast or baked potatoes left. There was chocolate cake for dessert and pistachio ice cream to go with it, and Mark Morley put on a pot of fresh coffee. He cleared the table, scraped the plates and rinsed the cups and mugs in the sink, and finally started soaping the dishes with a sponge and wiping down the utensils.

"Good coffee," Bob Doyle said.

"Thanks. I don't drink the stuff myself, but I like the way it smells."

"That roast was excellent."

"Glad you liked it. When you move around a lot, you cook or you starve."

"True," DeCapua said.

Morley was drying a fistful of forks. "We'll be heading back out in the morning, so make sure you all get a good night's sleep."

"What time you thinking of leaving?" Mork asked him.

"Late morning."

"What's the forecast?"

"Right now the radio is saying seas about five to ten feet, and winds about twenty knots outside. Just the window we're looking for."

DeCapua said, "On the TV they're saying a blow is coming this way. A good one. It's south and east of the Aleutians right now, moving fast toward us."

Morley put the utensils in a drawer.

"Well, we got to get the gear."

"Why?"

"Because otherwise it's lost gear."

"Yeah," DeCapua said, "but it's not your gear, right?"

"No."

"Well, then, what difference does it make? I mean we're bringing the boat back. We've already not made money. So what if we lose the gear?"

Morley looked at him appraisingly. "I want to get the gear back."

No one said anything.

"Besides, I've talked to the owner," Morley said, "and he knows we're going out to get the gear. Now, while we're out on the grounds, maybe we can throw out a couple of quick fives and let them fish while we're retrieving the gear."

DeCapua sat back.

"Before we spot the gear, we can throw our shit out—maybe two sets of five skates—real quick. We let the new stuff fish while we get the gear we've lost—buoy balls, anchor and all. How many anchors are we down?"

"Two," Mork said.

"*Two?*" Morley scowled. "Those things are a hundred bucks apiece. Do we have enough for a fifteen-skater?"

"For that, we'd be short an anchor."

"Christ."

"We got an old car battery down below," Mork said. "Maybe we could tie that off to hold a string."

Mike DeCapua sniffed. "Say," he said, "isn't there any other place to fish along the coast?"

"Like where?"

"Like up past Graves Harbor. We could go pull our gear and do a bunch of smaller strings inside."

"Well," Mork said.

"A lot of people fish that area out of Yakutat. You ever fish up there, Giggy?"

"Yeah."

"There's Laykee Bay up off Yakutat."

"I know it."

"Okay, look. Say we get out to Fairweather, you know, we get our gear, we put, say, two strings out, we get nothing. Here's another option that we can plan on doing."

"It's an idea," Mork said, but he did not sound convinced.

"Why get knocked around out there if we can catch smaller, specialty rockfish in the inland waters? So we don't get lots of big fish. We can get specialty rockfish. They're worth more."

"What kind of price you hear on yellow eye today?" Mork asked Morley.

"Owner said a buck thirty-five a pound," Morley said. "That's for number one product. More for the specialty rockfish."

"How much more?"

"Maybe fifty cents a pound more."

"See?" DeCapua said. "We get a bunch of the little guys, on the inland waters. And what about sand sharks? You guys told me when I came on this boat that we were going after sharks. Hell, we've thrown at least two hundred sharks back. Ten percent of that is keeper. What do you think, Dave?"

David Hanlon had been listening intently but had not said a word throughout dinner.

There was a long pause.

"I'd rather not get knocked around any more than I need to." He lifted his shoulders and let them fall. "Maybe it's not a good idea to go out."

A stone silence followed.

"Well," Morley said softly, "we're here to fish." Hanlon looked down at his glass. Morley went on: "Anyone else got something they want to say about this?" He paused. "Everyone's tired. I'm tired, too. But we got a chance to make money here. Let's hump and get our pounds and get the hell home."

Nobody said anything.

"Hey, Bob," Morley said.

"Yeah?"

"You call your kids?"

Bob Doyle nodded. "Sure. Sure did. They're good. I spoke with my baby girl, Katie, tonight."

"Oh yeah? What did she say?"

"She said she wanted to hear another Barbie-as-Fisherwoman story. I always call and tell her a story about Barbie's adventures. It's a little thing we do."

Bob Doyle felt his face reddening.

DeCapua laughed. "You don't want to tell her anything about the Barbie I once knew."

It was late now and Bob Doyle had been unable to sleep. Once he had gone out on deck and looked up at the sky. It had cleared and the wind was high up and the moonlight lay on the branches of the snow-heavy spruce. He lit a cigarette, looked for a place to dispose of the match and finally tossed it overboard. A deer was picking its way along a trail. He watched it for a minute and then got bored with the deer and went back to his room.

Mike DeCapua was still up. He had found a paperback after dinner on the borrower shelf of the community hall, *Civil War Chronicles*. He hadn't put the book down since. He's some reader, Bob Doyle thought. How is it that Mike has no patience for certain things and yet won't put down a novel as thick as a phone book until he's read the last page? He thinks he knows everything and he does know quite a bit about a lot of things. But he's also got that streak in him. I don't know where he gets it but I've seen that streak in other people before and it never did them any good.

Mother had that streak, he thought. Deep down she had a good heart but she had that thing in her, too. Dorothy May Harlow. She was a direct descendant of the first people to come to America on the *Mayflower*. It made for a cherished fable. Truly, he thought, how many people could claim to be a *Mayflower* child? But she was. And she had that streak. She had it and he supposed he had a little of it, too. Those kinds of qualities were passed on. Certain things you carry around inside of you from your parents and grandparents and great-grandparents. No matter whether you want them or not, you carry them.

So he had things from Dot Harlow. Some dancer, she was. When she was young she danced every Saturday night at the Elks Club. That's how she'd met Dad. That's how she'd met Uncle Jim. Dancing at the Elks. She wound up marrying Ed Doyle, of course, and they had his older brother, Jim, before he went into the navy. Then his father had come back from the war and they had another boy, Dick. Then Dot decided to divorce his father and marry Uncle Jim.

Must have been hell on his dad, he thought. But he just moved back to the Doyles' home place across the river in Walpole, New Hampshire. And then Mother and Uncle Jim bought a bright yellow house with ten rooms in Bellows Falls. They tried to have a baby. They tried and tried. After a while they did not try as often. Uncle Jim drank a lot. It must have bothered him something awful to see his brother, Ed, moping alone under the maple tree out front of their parents' old clapboard house. Uncle Jim would return home drunk and holler at Dot and break things in their big, new house. Dot would clean the mess up, always without a word.

One year Dot went to a missionary priest who was visiting the St. Charles Church and cried and told him she loved Uncle Jim, not Ed. "Go back to Ed," the priest said. "The Lord will smile on you if you tell him you want him back." So Ed and Uncle Jim switched houses.

Right away, Dot and Ed had a third son—himself—and afterward his sister, Sally. Uncle Jim soon died from cancer. Then his mother and father divorced for good. So he and his younger sister, Sally, got raised in the bright yellow house in Bellows Falls. Later, his mother married a man named Fred Perry and stayed married to him for twenty-eight years. They became members of the Bellows Falls Moose Lodge and danced often at parties and went bowling and made each other quite happy.

All that time his father laid track for the B&M railroad, sorted mail at the post office, wrote poems about sunsets, drank heavily and sat alone at mass on Sundays in the back pew of the St. Charles Church.

Lying there in his bunk, Bob Doyle remembered the last time he saw his father. His old man had been hospitalized with cancer. So he took a leave from the Coast Guard and visited him. They played cribbage. He beat his father over and over, razzing him, devilishly, as he always had. A few weeks later, after his ship had left port, he was handed a cablegram informing him of his father's passing. For years he felt terrible that he had beaten his father and made fun of his cribbage game that way, and he was unable to blunt the guilt with anything but alcohol. It still bothered him, he supposed. He had never forgiven himself for it.

And what about that last day of seventh grade when he and Ronnie Hamilton and Dickie Keating decided to go skinny-dipping at Twin Falls? How they were laughing and walking along the country road, the dirt soft in the shade of the cottonwoods and pines, the smell of the hot sun in the trees, he and Ronnie and Dickie, a whole summer vacation ahead of them, each chewing cigars that Dickie had nicked from his dad's box of Jamaicans, when a little blond girl, all wet and out of breath, jumped out of the bushes in front of them, tears hot and dripping off her chin.

Can you swim?

Why?

My brother is sleeping under the water.

He was never able to recall what became of his cigar. All that came back clearly was him slipping and sliding down the steep, rocky slope through the shadowy light of the pines and nearly skidding off a ledge in his haste, a ledge that jutted twenty feet above the prongs of gray rocks at the toes of the falls. He heard a man and woman shouting the name Carl and then spotted the couple in sopping underwear crawling along the rocky banks toward a pool of water at the bottom of the falls. And then, gazing straight down, horrified, he took a few steps back and then ran forward and leaped, arms and legs flailing, and hit the water with a heavy splash.

The water was very cold. He stayed under as long as he could without cry-ing out. Reaching through the bubbles, he touched something clammy and opened his eyes and saw the white, white face and the open eyes and he real-ized that he had the hand and he started to pull and pull and pull until his lungs hurt and he let go and came up and took a breath and went down again. This time he grabbed under the armpits and pushed off the bottom with his feet and they were going up. His head broke the surface and he could see the banks and he thrashed and fought through the water and then dragged the boy ashore and laid him gently down on the rocks.

The boy's face was swollen and pulpy, the hair black with bits of pine needles smeared over the eyes. He lifted the head and, brushing the hair back, saw the flat, shiny eyes and let go of the head. When it cracked on the rock it did not even jerk.

"You awake?"

"Sure."

He opened his eyes and saw the silhouette of the bunk above him lit around the edges by the glow of the reading lamp. He turned his head. His children smiled at him from two pictures taped to the wall.

DeCapua said, "It looks like we're going to go fishing those grounds again."

"Think so?"

"That's why he's got us fixing all that shitty gear. We wouldn't need it otherwise." Bob Doyle was half listening. He was lying still, studying the pictures.

"It would be stupid to go all the way out there, seventy miles, just to pick up gear," DeCapua went on. "He wants to fish, all right."

"I guess so."

"I'm turning off the light."

"All right."

When I die, Bob Doyle thought, I wonder how much of me my kids will carry around inside of them. Whatever they get from me, I hope it helps them. It's too late to change things, anyhow. Or is it? He was going to sleep a little while. He was tired. Too tired.

TWENTY-TWO

Bob Doyle woke before the others and ate a bowl of oatmeal alone in the galley. Then he fixed himself a coffee, black, and took it out on deck. The sun was behind the clouds and the wind was blowing. High on the tops of the mountains the big spruce trees were swaying. He turned an empty crate over and sat on it. As he was listening to the water on the rocks his hand trembled and the coffee began spilling on the deck. He put the cup down and opened and closed his fingers. Then he shook his hands, rubbed them together and drew a cigarette from a pack he kept in his shirt pocket. He lit it and, leaving the cigarette in his mouth, blew the smoke through his nose. It was all right like that.

He was smoking his second cigarette when he heard someone stirring in the galley. David Hanlon came out. He smiled and went to the bait shed. Bob Doyle picked up his mug and went inside. He heard Gig Mork and Mark Morley talking in the pilothouse. He also heard the radio. The forecaster was saying a low-pressure system had spun off from an enormous Aleutian low and was tracking north and east. It would be in the eastern gulf within forty-eight hours. Seas were expected to build to thirty feet, and a gale warning was in effect for late Friday night.

"Hey, Bob," Mork called down to him. "Do me a favor and make sure there's enough gas in the generator. After that come out on deck and help me fix the last of that snap gear."

"Sure."

Morley came down the stairs and told him to be ready to shove off at noon.

"That's fine by me," Bob Doyle said. "Looks like we've got a great day ahead of us."

"You eat?"

"Yeah."

"I'm making some eggs and sausage, if you want some."

"No, thanks."

"Where's that DeCapua?"

"Sleeping."

"Well, get him up."

After Morley had his breakfast and went back upstairs, Mike DeCapua came out. He had a bad morning hack. His eyes were red and the skin around the eyes was swollen. He shot Bob Doyle a sharp glare.

"Why the fuck you look so happy?"

"I don't know," Bob Doyle said. "The nice weather, I guess."

"Oh, is it?"

Mike DeCapua went to the door and frowned.

"It's too fucking nice," he said.

Just before six that evening Gig Mork came down from the pilothouse and tapped Mike DeCapua on the shoulder. They had been running a straight slot to the westward with the seas to port. There was no wind to speak of and the *La Conte* was making eleven knots with little effort.

"Your watch," Mork told him.

"Aye, aye, Captain."

"Cut the crap."

It was dark now and the boat was steaming through heavy fog. There was maybe two hundred yards of visibility—no more, even when they turned on the searchlight atop the wheelhouse.

DeCapua took his seat behind the wheel and glanced at the smoky mist curling and clouding the windows. Behind, in the small cabin, he heard Mark Morley rolling in his bunk and groaning.

"So," DeCapua said, and he cleared his throat. "What's with the chums, skipper?"

"What?"

"The bait. Why are we using chums for bait?"

"Well," Morley said, "we got them for a good price."

"I guess you must know that it takes a lot longer to bait with this stuff," DeCapua said. "Everything to do with this bait is harder. That's why I'm asking."

"Bait is bait."

The bilge alarm sounded and flashed. Morley said, "Turn that damn thing off. It's been doing that ever since we left Elfin Cove. Turn it off."

DeCapua flipped a switch near the ceiling and the light went out. Then he sat down and made a cigarette, wetting the rolling paper with the tip of his tongue. Then he lit it.

The engines were running smoothly and water was racing past the hull of the boat.

"So you been to the joint?" he asked Morley.

"Uh-huh."

"Me, too," DeCapua said.

"How long you been on parole?"

"Long as I can remember."

Morley said, "I was wondering, you know, if you think we're violating parole right now—being on this boat together and all?"

"No," DeCapua said. "I fished with a guy who was on parole once, and it didn't matter. The PO said we were on the job, so it was okay. If we'd been hanging out together drinking and shit, there'd be a problem. But when we're on the boat doing boat stuff, that's not hanging out."

"Oh."

"When we get in town, we can't socialize."

"No."

"Say, skipper?"

"Yeah?"

"What you say we just get this gear and turn around and go back inside? I mean, I know you want to fish out here and all, but we could do just as good, say, up near Yakutat or Graves."

He turned and saw Morley lying in his bunk, face to the wall. The skipper did not move.

"Let's do the math for a minute," DeCapua said.

"Math?"

"Yeah, let's figure out what we can catch and what we'll make and you'll see what I mean."

The skipper was still looking at the wall.

"This isn't about math," Morley said.

Gig Mork handed the binoculars to Bob Doyle and whistled.

"You give it a look."

Bob Doyle swept the ocean with the big glasses. It was drizzling and a pall of fog hung all around them. The back of his neck felt like a piece of ice.

He shivered.

"Well?"

"Nothing."

They were standing on the raised platform around the wheelhouse looking off to port where the spotlight pried into the blackness. The engine was almost at an idle. For a half hour they had been crawling back and forth across the patch of the Triple Forties, where, according to the video plotter, they had set the two lines of gear.

It is as spooky as a graveyard out here, Bob Doyle thought. Maybe it's just the calm. This sea acted so wickedly just a few days ago and now she's so gentle, so kind. It's as if she's saying that she's sorry for what she did the last time and that the wickedness is gone and we will be friends again. But I don't believe any of it.

He tried to scan the ocean through the fog but there was no sign of the buoy. Perhaps it snapped off the main line and floated away, he thought. Or maybe the coordinates on the GPS are wrong. If the coordinates are far off it is going to take us hours to locate that lousy buoy marker. And who knows how much line is still attached to it?

Just then the searchlight caught a speck of orange no bigger than a pinhead. Bob Doyle focused the glasses.

"I see something," he said.

They swung the boat around and motored over through the light chop. It was the marker, all right. Morley cut the motors down to an idle and they coasted alongside of it.

Mork leaned over the railing and hooked the buoy with the gaff. Morley came down from the pilothouse.

"Is it still hooked on?"

"Shit yeah."

"Well, let's get it up."

They set up on deck as they had before—Morley at the winch, DeCapua on a crate in the middle of the foredeck, Bob Doyle near the hold hatch standing by with his knife. The main line started coming out of the water.

"Jesus," Morley said.

Only skeletons dangled from the hooks. There was not a shred of skin, not even cartilage. They did pull in one fish with some flesh on it, a four-foot halibut It was pecked to pieces, though. Rotting. The eyes were gone.

Morley knocked the skeletons off with the gaff. They made a little crack when he hit them, like a bird's neck snapping.

The bones that fell on the deck Bob Doyle tossed over the side. DeCapua coiled up the line. He did not say anything or look up. David Hanlon took the coils to the bait shed.

When all the line was in, Morley tossed the gaff down and yanked off his gloves.

"At least we got the gear back," he said.

In the heavy calm before the wind rose they set out the two strings of old gear they had been able to retool in Elfin Cove. Each string had five skates, not quite a mile of longline. Morley told them he wanted to put out short strings so that they could haul them in quickly, just in case the storm came faster than expected. They were short an anchor. Mork went below to the engine room and came back with the old car battery and tied it off to the middle of the line.

By one in the morning the strings they had hauled back had been rebaited; they also had two other skates ready to go, a box of bait thawing in the shed, the bilges pumped and the generator topped off with gas. The ice was waiting in the holds.

They all filed into the galley, peeled off their raingear and sat down around the table. Bob Doyle peeled and ate a banana. Mork put on a pot of coffee. DeCapua rolled his cigarettes. Hanlon drummed his fingers on the table, looked out the glass door.

They were afraid to jinx the sets. Nobody dared say a word until Mork broke the silence.

"Hope they're fishing good."

"Really."

"Uh-huh."

"Think we set them in the right spot?"

"Hard to say."

They all went quiet again.

"What time is it?"

"Almost three."

"Anyone want more coffee?"

"No."

"No, thanks."

"Bob?"

"No, no."

"We must be getting close to where we set them."

"What do you think?"

"I say we get out there and get them."

"Okay."

They all stood up and tightened their raingear. Mork put his coffee cup back in the sink.

"Hell," he said. "Those goddamned lines better be fishing good."

Actually, they could not have fished better. Not a single hook was empty. They caught yellow eye, tigers, calico, lingcod, halibut and the occasional sand shark, too. Morley had to slow the winch—sometimes almost to a stop—there were so many. He would reach down and sink the gaff hook into a fish and lift it up and back over his shoulder and then turn and see five more lifting out of the sea. They were no small fries, either; Morley had to use two hands on the bull hook to tug the fifty-pounders in. The yellow eye came up, gills flashing, their lean, orange bodies bright in the glow of deck lights.

"Here they are, boys!"

DeCapua, coiling now like a machine, yelled back: "Well, it's about fucking time!"

"Hell, yes!"

Morley was rapping the gaff on the hull he was so excited.

"Oh, Mama! Lookie here! We got a tiger!"

He lunged and sank the bull hook into the fish, dug his boots into the no-skids and yanked the yellow eye free of the hook and up and over the railing. It landed on the deck with a heavy thud.

"There's a forty-dollar one!"

Bob Doyle skipped over, lifted the gills and jammed the blade in deep. The fish shuddered.

"She's a beauty!"

"Coil coming!"

"Ooh, ooh! Look at the size of this son of a bitch! Just look at him!"

"Yeee-haaa!"

Just then the lingcod broke the water. It was long and round as a saw log and it writhed and twisted as though it wanted to take Morley back into the sea with it. Bob Doyle marveled at the fight in the fish; he could see it was hooked in the side of its mouth and still it was writhing, trying to wriggle free. It was one fine lingcod, all right.

Two hands gripping the gaff, his muscles rippling taut and burning, Morley lugged the wriggling fish in and hurled it on the foredeck.

"Gut him, Bob!" he yelled. "Stick the son of a bitch!"

Bob Doyle had one arm and a knee on the fish, and was trying to go for its gills, trying to get the blade in the head, missing, stabbing again, missing again as the flailing, squirming lingcod wrenched at his arm. Then he felt a quick cutting and the skin across the back of his hand open.

"Cocksucker!"

"Stick him!" Morley shouted. "Stick him!"

Mike DeCapua was in hysterics. *"Ha! Ha! Ha!* He's beating you, Bob!"

"Shut up!"

"Watch out for his mouth, Bob! Those teeth are sharp!"

The fight dragged on. Bob Doyle pressed his knee into the wriggling, snapping fish. As he tried to jam the blade into the fish's head, the lingcod bit him.

"Bastard!"

"Quit fucking around, Bob!" Morley shouted. DeCapua had not stopped laughing.

"Shut up, Mike!"

Mangled, its guts spilling out, the cod thumped one last time and went still. Bob Doyle collapsed, sucking air.

"Gee," DeCapua said, catching his breath, "that was pretty, Bob."

"Fuck you."

"Aw, don't get sore," DeCapua said, and he laughed until he coughed. "There's plenty more."

There was, too. Sometimes Morley did not hit a fish right and it would slap back into the water and float with its bladder out. David Hanlon would lean over the railing and scoop the dying fish out of the sea with the other bull hook. Later, while they were finishing the first string, another heavy lingcod ripped the aluminum gaff right out of Morley's hands. The fish and gaff went cleanly into the sea. It made Morley sore; from then on he had to use the heavier, wood-handled gaff.

Then seagulls started attacking the catch. They went for the eyes and Morley swore at them and tried to scare them off with the bull hook, but they only flapped their wings and hovered just beyond his reach, squawking, and then attacked as soon as he went back to operating the winch.

All the while the breeze was stiffening and the sea building. And the fish kept right on coming.

"This is fucking crazy," DeCapua said. "I'll bet we got more than a ton on these strings."

"Those fish bins hold eight hundred pounds," Morley said. "How is it, Bob?"

"I'm almost filling the second bucket now."

"Damn," DeCapua said.

They worked at a breakneck pace: Morley gaffing and hauling without letup, hooting and dancing a jig and chanting, "Hey, ho, up she rises"; Bob Doyle, stabbing and kicking fish into the bins, his arms and hands stinging from the slashes left by the lingcod; DeCapua, coiling in the same mad, fluid motion, shrieking with laughter when a lingcod bit his partner; Hanlon, scampering the length of the ship, gathering line and running it back to the shed to clean and bait it; Mork, working the boat into the current, backing her down to ease the drag on the line, jogging her in the swells, which were considerably higher now than they had been when they first set the gear a couple of hours earlier.

"Hey, Giggy!" Morley called up topside. He held up a yellow eye, a thirty-pounder. "Look here!"

"This is better than sex!"

"Hell, yeah!"

"This fucker is finally paying off."

"*Fucking A!*"

Once all the skates were up, they stuffed ice into the mouths of the fish, iced them in the bins, then sealed the totes and lowered them into the holds. Then it was back to fixing the broken snaps and gangions, scraping bait hooks, before setting out again.

"This," DeCapua told Bob Doyle in the shed, "is how the *Min E* could have produced, Bob, if Phil had ever decided to bust his ass."

"Think so?"

"Oh, I *know* so. That lazy prick."

"Old Mike."

By six-thirty that Friday morning, the thirtieth of January, they were hauling the second pair of strings. These had even more fish than the first two skates and there were fewer sand sharks and halibut to throw back. They filled two more totes, solid, with yellow eye and topped off their second tote of bycatch, which met their quota. There was no more need to bring aboard the lingcod.

"Breaks my heart," DeCapua said.

"Go to hell," Bob Doyle said.

The skipper had planned a bacon-and-eggs and hash-brown breakfast for his crew. But the stove went out again. There was no time for bitching; they munched on peanut-butter-and-jelly sandwiches, bananas, baloney and mayo on English muffins, and went right back out.

All this time, the seas were picking up. The skies were unloading a driving gray rain that raked them like buckshot, and what had earlier been a soft breeze was now a stiffening southwesterly greater than ten knots, with the occasional odd gust from the east.

But the fish kept coming.

By daybreak, they were lowering into the holds three more totes crammed with yellow eye. This brought their total catch to five thousand pounds. The extra weight made the stern sit even lower in the water. Even the medium-size curlers began bursting on the railings and lifting sheets of spray over the rear deck. It was streaming out the scuppers, but more water was finding its way into the engine room.

The two bilge pumps down below were not keeping up with it, so Gig Mork had Bob Doyle and Mike DeCapua help him lug the backup generator down below. They set it up on a wood pallet beneath the stairs, attached one end of a hose to a suction fitting, ran the other end up and out the hatch and over the side railing and began using the generator to pump the bilges.

Later, they were in the galley, smoking. Morley came downstairs and asked about the generator.

"It's helping," DeCapua said, "but we're gonna have to keep an eye on the water down there."

"It's supposed to get snottier," Morley said. "They're calling for twenty-footers later this afternoon and winds up to thirty knots. What I'm thinking is to put out one big string right now—a twenty-five-skater—and pull it before it gets really shitty."

"How big?"

"I said twenty-five."

That was more than five miles of longline. To set it would take at least ninety minutes; to haul it back would be another three hours.

"Dave tells me we got thirty skates ready," Morley continued. "So, what we'll do is set out a twenty-five-skater, then cut the motors and drift right off the end of the line. We'll wait an hour, then haul the sucker and scoot."

"I don't know," DeCapua said.

"Giggy likes the idea," Morley said. "What you think, Bob?"

"Sure."

DeCapua grunted.

T W E N T Y - F O U R

It was rough setting out the big string. They kept going weightless between the big swells and it was hard to hold their footing even inside the bait shed. The baiting went slowly. David Hanlon did not look good. His face had gone pale as scraped bone, his eyes barren and rimmed a dusky red, and a number of times he dropped the line he was working on and ran outside to get sick. Now the waves came arching over the bow railings, thudding on the deck with a hard, white burst, and swamping the decks in a broth that froze their legs up to the thigh. With the larger swells the fantail lifted clear of the water and they heard a high-pitched, unsettling screech of the driveshaft and felt the breath-stopping emptiness of sudden weightlessness. Each second they went weightless was a lost second for setting gear; the knots were rushed, the skates tangling and snarling in the chute. A lot of the hooks hit the water without any bait. But they did get the entire main line out.

Once it was out, Mark Morley said to Gig Mork, "Okay, let that sucker fish for an hour, but no longer."

"All right."

"I'm gonna lie down. I'm whipped. Take the helm for me. You know how it works. Stay close to the buoys and get me up when we're ready to pull."

They looped around the set, circling the orange buoy markers, riding in the belly of eighteen-foot swells. Mork would point the *La Conte*'s nose into the oncoming swells and jog her for a while, and then swing the boat around and follow the seas. David Hanlon never left the bait shed, except to get sick, and Bob Doyle and Mike DeCapua took care of clearing the decks and lashing down the pallets, buckets and loose fuel jugs. They tried to keep focused on their jobs, but every so often Bob Doyle felt his gaze drawn overboard, to the ocean building around them.

At noon, Mork woke Morley and the skipper came down to the foredeck and manned the winch. At first, things looked promising. Their first several skates had a decent number of yellow eye and a few sand sharks. But after the fifth skate the numbers dropped off and they pulled more snarled, knotted balls of line from the water than anything else.

"Storm's coming fast," Morley shouted to DeCapua over the shrieking wind.

DeCapua shouted back, "No kidding."

Just then a terrific wave rose up off the bow. It towered over the railing like a huge oak tree, hanging for a fraction of a second, and then fell forward in a gray-black roar that made the deck roll under their knees. Morley crumpled under it and for the longest moment, Bob Doyle lost sight of him. Then the deck waved up again and Bob Doyle saw the burly back lifting through the foamy water like a surfacing sub.

He sloshed his way over toward the skipper and steadied him by the elbow.

"You okay?"

Morley could only nod, pull off his glasses and wipe the brine from his eyes.

Rushing, they pulled the last ten skates without a hitch and brought on several hundred more pounds of yellow eye. Now hail the size of quail eggs was coming with sheets of rain, the wind slinging it all into their faces.

Mike DeCapua had been coiling quietly, precisely, head down to keep the hail out of his eyes, butt stuck to the wood crate on which he crouched despite the freezing seawater, which was now above his waist. When he wrapped up his last skate and shoved the coil along the deck to Bob Doyle, he took a moment to look out over the port railing.

The barrels of the waves were big enough to swallow a house. But what startled him was the line—a line darker than ash that stretched along the length of the western horizon.

He struggled to his feet.

"Skipper!"

He stumbled over to Mark Morley and grabbed the skipper's arm. "Hey!" Morley, who was sweeping fish into the holds, did not look up.

"Hey!" DeCapua shouted. "We got to get out of here!"

"Why?"

"That's why."

He pointed to the horizon. The black line now was twice as thick.

"So?"

"So? *So?* Man, in Alaska if you see a line like that out there, you get the fuck off the water!"

Morley hadn't stopped tossing fish in the bins.

"Hey! Hey!" DeCapua snapped. "Listen to me. I ain't shitting you. That line is going to be on us in an hour."

Morley gave the horizon a fast glance. "You ever been in a storm before?"

"Yeah."

Morley leaned over and grabbed a yellow eye by the gills. "It ain't gonna be any worse than around Coronation."

"This ain't Coronation."

"Hey, Bob," Morley said. Bob Doyle turned around. "Go back aft and ask Hanlon how long it'll be to get another string ready."

"Hey, I don't think you heard me," DeCapua said to Morley. Bob Doyle had gone to get the gear. "Tell me you're shitting me."

"No."

"Listen," DeCapua said. "We've lost almost all our anchors. We're eighty frickin' miles from the beach. It's gonna take us an hour to just throw out a string. Then we gotta let it soak. That's another hour. Then we gotta pick it up—"

"What's your point?"

"Why don't we check the weather before setting out another one?"

Morley glared at DeCapua with his mouth closed so that his lips made a tight line, like the mouth of a fish. He looked to westward and sighed.

"All right. Wait here."

He went inside and up into the wheelhouse. Bob Doyle came back. "Hanlon says we can get a ten-string up in less than an hour," he said.

"Shut up."

Just then a wave pummeled the bow and the deck dropped off and away from under them. They threw themselves down on their stomachs and grabbed the hatch to keep from being washed off. The boat heaved, wallowed and finally stabilized. Morley came back down.

"Say, Bob," he shouted so as to be heard over the wind. "How much fish we pull on that last string?"

"Enough to fill a tote, maybe two."

"All right," Morley said. "You win, Mike. We're going back. No sense getting beat up and losing gear for another tote of fish."

"Hallelujah."

"What we got is respectable. Get these decks tightened up. Get this shit iced. Afterward, you guys hit the rack."

They wound up leaving the catch out on the foredeck. It was too risky to be operating the cranes with the boat pitching and rocking so violently. Mork and Hanlon lashed the fish down with netting and threw ice on it while Bob Doyle and DeCapua stowed the gear in the shed.

"It's about fucking time we got out of here," DeCapua said. "I swear to Jesus that guy's got a fucking death wish."

Gig Mork threw open the bait-shed door.

"What you say?"

DeCapua turned around. "What?"

"You lousy son of a bitch. I told you to shut the fuck up. Know what that means? Shut the fuck up?"

"Hey, take it easy, Giggy."

"I hear any more out of you, you fuck, and I'll kill you. I'll kill you right here and now."

"Take it easy."

"Bastard," Mork said. His eyes were wild. DeCapua looked away from them. Mork turned to Bob Doyle.

"Stow all that gear under there. Tie it up good. And get all those hooks and snaps down off the wall."

He stormed out.

They checked the scuppers and dogged down the aft hatches, fuel jugs, lockboxes, gaffs, batteries, hydraulic hoses, clamps, pallets and pumps. They shined a flash in the bilges. The water was up almost two feet and was sloshing just six inches below the base of the engine. Even with the two bilge pumps and the backup generator running at full tilt, it took them more than ten minutes to get the water out.

Mark Morley found them sitting around the galley table, resting. He had come down from the wheelhouse to give them their watch assignments.

"From now on, I want someone checking those bilges every half hour," Morley said. "And don't forget, when it's your turn to go below, first you tell us you're going and you don't go without your partner. Understood?"

They all nodded.

"Mike, I want you to take first bilge watch."

"I'll do a double," DeCapua said.

"All right," Morley said. "Then you'll be on until four o'clock. Bob, Dave, you guys work out between you who goes next. But I want somebody down there checking those bilges every thirty minutes. If you see something wrong, come get me or Gig."

They nodded.

"Mike, come on down with me. I want to show you some things down in the engine room."

When they left Hanlon looked at Bob Doyle. "You look tired, Bob. I'll take the second shift."

"Sure?"

"Yeah. Where's the coffee?"

"The grinds are in that cabinet, there."

Just as David Hanlon stood up a wave hit and the galley tilted sideways. The cabinet doors flew open and plastic cups and books tumbled off the shelves and slid across the galley floor. Bob Doyle grabbed the dinner table to keep from being lifted out of his seat. He looked up and saw his partner in a heap on the floor.

"Hey," he said. "You okay?"

Hanlon stood up.

"Fine," he said.

T W E N T Y - F I V E

He felt a hand on his shoulder. It was David Hanlon and he said, "Time to get up."

Bob Doyle opened his eyes but the rest of him was still asleep.

"Time to get up."

"All right."

He wanted to move but he felt so heavy. Hanlon shook him again.

"All *right*," Bob Doyle snapped. The heaviness had not gone away but he was able to pull his legs out of the sleeping bag. Hanlon watched him for a moment and went out.

The room kept tilting way over to one side, then to the other. Slipping a foot into his boot Bob Doyle felt a force pull him backward and he banged his head on the bulkhead.

Mike DeCapua came in. He was wet to the skin and dripping.

"What's wrong?"

DeCapua grunted and climbed into his rack. "We're taking on a bunch of water. Pumping every half hour ain't gonna be enough."

"No?"

"Just make sure you get down there and run that fucker every twenty minutes. We can't lose that generator pump. That's the only pump worth a shit that we got."

"Right."

Hanlon was bundling himself up in the galley. He looked pretty bad. He hadn't looked good since the seas kicked up during that second set; but he looked pretty bad now.

"Hey, Giggy," Bob Doyle called up the companionway. "Me and Dave are going out now."

"Go!"

He nodded to Hanlon and reached for the door.

The gust cuffed his face as soon as he stepped outside. He felt his cheek. It stung. Some wind, Bob Doyle thought. He slid the door shut and tried to latch it. But his hands were shaking.

"It's still open," David Hanlon said.

A swell toppled just short of the railing and threw up a column of white water. Then there was one they did not hear coming until the whooshing cough. They both went flat against the door and with the bump and burst of a wave on the foredeck heard the rattle of falling clods of spray. Bob Doyle took a step back and tried the latch again.

"Hurry."

"I got it," he muttered.

Just then he heard above him a heavy breath like a hiss, then a rushing thunder—then there was the shock, as if a gaff had come down on his skull, and the unbelievable heaviness and Bob Doyle knew the wave had him. He tried to breathe but his breath did not come and he felt himself going flat and everything around him darkly cold and heavy and he had a thought that he would never move again. He gave up fighting the weight and let himself go, knowing that he was going overboard and feeling as though he was inside nothing at all, but instead of sliding away and float- ing he felt his cheek pressing hard on a rubber pad and he realized that he lay sprawled on the foredeck. He tried to move his legs and they moved a little, and then he felt something grabbing him by the armpits and lifting him.

It was Hanlon.

"*Move!*"

His legs felt watery. He tried to plant them firmly but they only bent and wobbled as Hanlon lugged him along the companionway. The next thing he knew he was sitting inside the bait shed, up to his butt in frigid water, his back against coils of line. He gasped for air.

"Breathe," Hanlon said.

He nodded.

"Breathe."

They opened the door a crack. The waves were coming every fifteen seconds. They burst white and flew in glittering fragments. Water was swirling and backing up around the scuppers.

They counted to three and dashed over to the hatch. It was open. They scuttled down the stairs. Steam and diesel exhaust stung their eyes. They heard the *chug-chug-chug-chug* of the engine, the jangling of tools and the panicky clatter of cans and tubes and chests. It was hot. Then Bob Doyle heard the sloshing. He put the beam of the flashlight between a gap in the floor planks.

Water was gathering below the engine.

Hanlon pulled open his slicker and collapsed on the bench behind the galley table.

"Take it easy," Bob Doyle told him. "I'll go tell Gig."

Gig Mork was at the helm, with the same stance and grip on the wheel that he had used to carry the steel anchors. His knuckles were white. The muscles in his jaw twitched.

"How is it down there?"

"All right, I guess."

"You guess?"

"We pumped out the bilges."

Sleet, hail and rain were coming in thick, steely braids, and curlers were pummeling the hull to port. Hills of water sneaked up on them from behind and arched and landed with such force that it felt as though a giant spike were being driven into the aft deck. When they could see such a wave breaking it seemed as though the *La Conte* was exploding.

"It's like a monsoon," Bob Doyle said.

"I been in worse storms in smaller boats," Mork said. "It'll be better once we get in closer to shore. We just got to keep jogging."

"How fast you jogging her?"

"Two knots."

"Where's Mark?"

"Asleep. How's Dave doing?"

"Not good."

They steamed up a steep, snarling swell. At the top the *La Conte* hesitated, then tipped forward into a void and plunged down its back side, down, down, down, bumping and slewing wildly as she went. It was as if they were strapped to a mare that was racing downhill into an unknown territory—a wasteland of dark, barren hills with no horizon in sight.

At the galley table David Hanlon sat with his head in his folded arms.

"How's the stomach?" Bob Doyle asked him. Hanlon didn't move. "There's Dramamine if you want it."

"No," Hanlon said. "That doesn't work for me."

A wave hit the bow. Bob Doyle took a step back, as though an invisible hand had shoved him.

"I don't want to be here," Hanlon said. "I wish I never came out." He lifted his head. His skin was white as an oyster. "It doesn't make any sense being out here in this."

"Don't talk like that."

"I can't help it. I've had a bad feeling all along about this. God, I wish I wasn't here."

"We're heading in now."

There was a noise like a gas tank exploding and the walls and floor shook again. The galley went a long way over and came slowly back. Hanlon peered at Bob Doyle, his eyes as empty as holes in a mask.

"I'm sorry I ever came out." He let out a thin, whisper of a sigh. "I shouldn't have come out."

"You've seen weather like this."

"Not like this."

"You're just tired. We're all tired."

"Never again. I'll never do this again."

"Well," Bob Doyle started to say, and all he could think of to add was, "That's good."

He reached up and opened a cabinet door. The cabinet was a mess inside. He rummaged around and found a can.

"Feel like some coffee?"

"No."

There was a filter in the drawer. Bob Doyle tossed out the old grinds and placed a filter into the coffeemaker. The galley tilted far over and dishes and pots rattled in the cupboards. A book slid off the shelf above the porthole.

"Sure you don't want some?" he asked Hanlon.

"No."

He spooned out fresh grounds and had picked the glass carafe up with his right hand when he heard a thudding burst and felt a great shudder and saw the door over him. He was lying on his back, covered in tins, plates, cups and loaves of bread. Before he could open his mouth he heard a loud yawn of wood and felt the boat tilting back and he tumbled end over end until his head hit something. He sat up. Plastic cups were floating around him. His head hurt. The carafe was still in his hand.

He got to his feet, dopily, and began picking up what had flown out of the drawers and cabinets. Then he returned the books to the shelves. The coffeemaker had not been damaged. Grounds lay scattered everywhere. Hanlon was curled up in the booth behind the table, his fingers clutching the tabletop.

"Look," Bob Doyle said to him, "in twenty minutes you can go wake Mike and get some sleep."

Hanlon just stared at him.

"Hey, maybe you ought to go now," Bob Doyle told him. "Why don't you do that? Go get some sleep."

Just then the floor leaped up. The coffeemaker flipped off the counter. As Bob Doyle lunged for it the carafe slipped from his fingers and smashed on the floor into many pieces.

He began picking out the glass shards from the sloshing water. Hanlon mumbled something and put his head in his arms.

By six o'clock the seas were twice as high as the ship. They rose up in huge dark walls now, their faces near vertical, thin, white lines across their brows. From the wheelhouse they almost could be mistaken for moonlit clouds.

"Light one of those cigarettes for me, will you?" Gig Mork asked.

Bob Doyle was sitting beside him on the floor of the pilothouse.

"Sure."

The boat was no longer clearing the tops of the swells; she was punching through the crests and launching out their far sides. Most of the afternoon the wind had been blowing largely in one direction, northeast, but now gusts were coming from all sides, in blasts, as if big shells were bursting around the ship.

"Dave's gone to bed," Bob Doyle said. "He wasn't doing too good so I sent him down to wake up Mike."

Mork nodded.

"As soon as butt-head is ready the two of you better go down and check on the bilges."

"Sure."

"Remember, nobody goes out alone."

"I know."

The lights dimmed, then quit. The computer screen went blank. Yellow, emergency lights flickered on.

"*Shit,*" Mork said. "The fucking laptop's out." He turned to Bob Doyle. "We're not getting any juice. Go down below and find out what's doing it."

In the galley, Mike DeCapua was pulling on his raingear.

"C'mon," Bob Doyle said.

"What's up?"

"The computer just went out. Gig thinks it's the generator."

"Glorious."

They timed the waves battering the hull, broke from behind the door and, heads bent and legs plunging, dashed to the stern. They knelt down beside the hatch and DeCapua yanked it open. Bob Doyle took one step down the ladder, and froze.

"Oh, God."

"What?"

"Jesus."

"Get out of the way."

The bottom step of the ladder was underwater, along with the generator and both bilge pumps. Water was rolling back and forth across the engine room and lapping at the motor.

"*Mama mia,*" DeCapua murmured.

"Get the skipper," Bob Doyle said, his voice cracking, "and tell him to activate the EPIRB."

The engine was still chugging, but water was halfway up its side.

What do you do? Bob Doyle was thinking to himself. What *is* there to do? Nothing. That's what. Don't do anything. Just wait for the others. They'll know what to do. But it's hell doing nothing. Jesus, it's hell. Let me think. There's got to be another pump. But what good would that do? We had three going and that wasn't enough to stop it. Where is it coming in? I can't see anything. Shit. All right, relax. Calm yourself down or you're going to start looking like him, up there.

He looked up the stairs. David Hanlon was staring back down at him. The whites of his eyes had devoured the irises. Stop it, Bob Doyle thought. *Stop looking at me like that.*

He listened to the *glug-glug-glug-glug* of the motor and watched the steam hissing from the block.

God, I could use a drink. Just one little drink. No, a big drink. A double. A double would do it. Well, you can forget that. Do something. Why don't you do something instead of wondering what the hell's going to happen?

Just then he heard boots on the stairs. Gig Mork came down and waded past him and the dead generator pump and over to the engine. He squinted at the temperature gauge. The needle on the dial was dropping.

"Goddammit."

Mark Morley stuck his head in the hatch.

"Holy Christ."

"The pump is gone," Mork called to him. "Forget the pump."

"What do we do?"

They started a bucket brigade; one man in the engine room, another on the stairs, a third at the hatch on deck. Mork started on the bottom.

"Give me that bucket! Hurry up! No time to fuck around now! Where the hell is Hanlon?"

Bob Doyle, who had just taken a full bucket of water from DeCapua and emptied it over the railing, said, "He's up here."

"Doing what?"

"Barfing his guts up."

Morley came running along the side of the boat. He threw himself down on the rolling deck. He'd been on the radio, putting out a Mayday.

"Any luck?"

"Who knows?" Morley answered. "I couldn't hear anybody." He lowered his voice. "I did set off the EPIRB, though."

"Which one?"

They had, Bob Doyle remembered, two Emergency Position Indicating Radio Beacons: one was a 406-megahertz model, the other a 121.5. The 121.5 sat in a holster in the wheelhouse. It was attached to a fifty-foot line. The 406 was a handheld beacon. If he remembered right, each EPIRB had a manual switch and a saltwater trigger, but the 406 emitted a stronger, more precise satellite signal. If anything is going to work in this, he thought, it will be the 406.

"The 121."

"What did you do with the 406?"

"Right here." Morley pulled it out of his rain jacket. It was the size and shape of a bowling pin.

"How long you think it'll take the Coast Guard to get here with a pump?"

So he thinks he can still save this thing? Bob Doyle thought. A pump. Oh sure. A C-130 might do a flyby, but the way this boat is rocking in these seas there is no way anyone's lowering a pump. Just getting a plane out here is asking a lot.

"It'll take them an hour."

"That long?"

"At least."

The boat was broadside to the waves now and taking wave punches up and down her portside. She was filling fast with water; each time she keeled and the water rolled in her belly she lost more of her center of gravity.

"Bob!" Mork called out. "Switch places with me."

"Coming down."

He bailed and bailed until he could feel his joints crack. The engine went right on thrumming. It was a powerful engine and hammered in perfect time. This boat isn't quitting easy, Bob Doyle thought. She is pretending none of this is happening. I wonder if it's correct to think of an engine as a she. Why not? She beats almost like a woman's heart. Feel the rhythm of her. I wish to Christ we could figure out where this water is coming in. It's cold as a son of a bitch.

He took an empty bucket from DeCapua and was dipping it again when he heard a sickening, gurgling gag.

He wheeled around and gazed at the engine.

"Holy Mary."

"Fuck me," DeCapua said.

They could only stand there, the two of them. The boat's heartbeat had stopped. The engine was dead. All they heard now was the maddening, high-pitched moan of wind in the rigging outside. To Bob Doyle it was as hollow a sound as a guttering candle being snuffed out.

By the time Bob Doyle retrieved his survival suit from his bunk and ran back to the bait shed, the others were already dressing up.

He yanked open the pouch and pulled off his boots. He didn't want the boots weighing him down in the water. He shoved his legs in the trouser legs, stood up, put his arms in the sleeves and zipped up the chest. Then he looked around.

"Where's Mark?"

"We thought he was with you," Gig Mork said. He was double-checking Mike DeCapua's zipper and lining, looking for holes or defects in the suit.

"He wasn't with me," Bob Doyle said.

David Hanlon was struggling to pull his zipper up. It was sticking just below the neck.

"Let me help you with that," Bob Doyle said.

"Do your own first," Hanlon said.

"Quiet. Let me help you."

"It's too tight," Hanlon said. "I can't get this zipper all the way up."

A wave slammed the stern and they collapsed like toothpicks. From his knees Bob Doyle saw the bait-shed door swing open. It was Morley. He was already wearing his survival suit. He had the 406 EPIRB strapped to his arm.

"Listen up!"

He told them they were going to bail again. They were also going to try one last time to figure out how water was getting in. If need be, they would dry out the generator pump. Mike DeCapua shook his head.

"What's your problem?" Mork asked him.

"That pump's gone," DeCapua said. "It ain't going to start."

"Maybe it will."

"It ain't working unless we take it apart and let it dry."

"Shut up, Mike!"

"Cut it out!" Morley snapped. "We ain't got time for that. Now let's get down to the engine room." He turned to Hanlon, who was still wrestling with his zipper, and handed him the 406 EPIRB.

"You hang on to this," Morley told him. "Hit the switch, right here, if anything goes wrong. But don't let go of it."

DeCapua just rolled his eyes.

The engine-room floor was under four feet of water but the lights were still on. Mork and Morley went back and forth along the bulkheads, feeling around for a hole, a loose plank, while Bob Doyle and the others bailed furiously.

In his survival suit, Bob Doyle was finding it difficult to keep his footing; the neoprene suits were buoyant and the water was rising fast. The more he bailed the more he felt the cold creeping up his chest. He handed a bucket to DeCapua and looked at the narrowing space between the water and the ceiling and was very afraid. Please, God, he thought. Get me out of here. He knew, however, that he would frighten the others if he said anything, so he kept his mouth shut and went on bailing.

Morley was ranting and swearing and pawing along the bulkheads. Mork took a deep breath and ducked under the water to check the floor for leaks. But he couldn't stay under long. The suit pulled him up.

"Find anything?"

"No."

"We gotta find it!"

"Mark," Mork said.

"We got to find where it's coming in."

"Mark?"

Morley turned around. Mork gave him a hard, long look. Morley sighed and lowered his head.

"All right," he said.

They regrouped on the foredeck in the lee of the pilothouse. The ship was lurching, listing so hard to starboard that at times the mast dipped into the waves. Through the drenching darkness Bob Doyle saw the waves rise darkly and break green and white in the rigging and each time the tumbling, boiling flood of whitewater cleared the deck, he noticed that several deck planks were missing. They crouched in three-point stances.

"Dave," DeCapua shouted. "Where's that fucking line?" Hanlon held up a roll of three-quarter-inch rope he'd grabbed from the bait shed.

"Give me that."

Bob Doyle leaned close to Morley and said, "Trigger that other EPIRB."

"Think so?"

"Do it now."

"Hey!" Morley shouted to Hanlon. "You got that EPIRB? Give it to me."

Morley took the 406 from Hanlon and held it up. Bob Doyle felt for the switch and pulled it. As soon as he released his hand, a powerful white flash blinded them.

"That's a helluva strobe!" Bob Doyle shouted at him. "Now they'll know nobody's out here kidding around."

A wave, this one straight up and down, rose high over the bow and hit the deck with such force that an anchor jumped out of one of the rollers. In the swirl of the water it flopped about like a speared halibut.

Just then a pallet ripped loose from the netting on the deck and came cartwheeling at them.

"*Giggy!*"

Before Mork could move the pallet clobbered him and he collapsed on his stomach. DeCapua crawled over to him. Mork was groaning and grabbing his head.

"You okay?"

"*No, I ain't!*"

DeCapua helped him sit up. Then he crawled back over and wrapped the rope around Hanlon's waist. He made a loop at the belt and tied a cat's-paw backed by a half-inch knot.

Then he grabbed Bob Doyle's arm. "How about some buoys?" he said. "We can tie them off to us. They'll help keep us afloat."

"Good idea."

Bob Doyle climbed the steel ladder to the top of the pilothouse and, grabbing the railing and hoisting himself up, swinging one leg over the bar, then the other, he dropped down into a crouch on the tar roof. He crouched low on one knee, never having felt so small, never having felt so fragile, the wind moaning like a prehistoric animal in his ears, sweat sliding down his flanks, and went to work on the knots that secured the buoy balls to the railing.

Now, for the first time Bob Doyle saw the storm in all its fury, ugliness and beauty: the towering, dark mountains of water, merging, pulling apart, bursting against one another; the tormented, black sky, an incessant discharge of electric hair, flashing like artillery; the speed-hardened wind, scooping out craters in the ocean, tearing the crests off the waves and carrying them into oblivion. He also saw the *La Conte* in all of her vulnerability: how she keeled in the troughs, her mast flailing, her rigging taking on a ghostly white coating from the geysers of spray that spouted off her port bow. And yet not a deck light had gone out. Even the searchlight was on.

The ship was taking knockdown punches, rolling and twisting under the combers, as though in agony, but refusing to go under. She's some boat, Bob Doyle thought to himself. But she won't last much longer. Ten minutes, if that.

He had the first knot undone and had started working to free up the second buoy.

Just be smart, he said to himself. Don't lose your cool and do something stupid. And be careful. This boat is lurching a lot. Don't relax and fall over and break something and wind up in the water alone.

There was a thickening lump in his throat and he tried to swallow but he could not. He kept working on the knot. Goddamn these mittens. How are you supposed to take apart a rope wearing a three-fingered, Gumby mitten? And goddamn this wind, too. All right. Relax, he told himself. Relax. I wonder if Sitka will send a bird out here? How long ago did Mark set off that 121? More than ten minutes. And now they had just triggered the 406. Just look at that strobe go. I guess we should have headed in earlier. I guess it was nuts all right to be fishing with something this bad coming at you. We shouldn't have tried it. Then again, look at all the money we caught. Shame we're going to lose it. I wonder how much our take would have been? Jesus, I want a drink. I guess I should have stuck to ordering supplies for ships. Blankets and chewing gum and cigarettes. That was a lot easier than this. Fuck boats. There's no money on boats anymore, anyway. Not for us little guys. Only for the big boat owners and their fucking fish permits. If this damned boat would only stop lurching I could get this knot. Christ. I should have got that job in the auto-supply store. Oh, sure. Order car parts for a living. Who drives in Alaska? Wouldn't hold it anyway. I'd drink myself out of it. Only place I don't drink is out here. I ought to live on the ocean. Why not? I'm going to die on it. Christ. Don't you *ever* get tired of self-pity? There. I got it. Two buoy balls coming right down.

Just then he heard a scream. He looked down over the railing. Someone was stretched out on the deck, writhing. He scrambled down the ladder, careful not to lose his grip on the lines attached to the buoys, and nearly threw his knee out landing on the heaving deck. He crawled over to Mike DeCapua.

DeCapua was clutching his crotch and upper thigh. A wave had ripped the lid of the hatch off and he had stepped into the hole. His leg had gone in all the way.

"It's me," Bob Doyle told him. "How bad is it?"

"Christ!"

"Hold still."

"Son of a *bitch!*"

Bob Doyle was checking DeCapua's suit for rips.

"Can you sit up?"

"Yeah."

He had not broken anything. The suit appeared to be all right. Bob Doyle looked at the wheelhouse.

He saw the strobe 121.5 EPIRB, still flashing in its holster inside the pilothouse. "I'm going to get that other EPIRB!" he shouted. "Get everyone tied together!"

He climbed the ladder, threw open the side door, grabbed the beacon and scuttled back down to the deck. The others were passing the rope, tying it around their waists and handing it off to the next man. Hanlon was on one end. Bob Doyle got on the other. He looped it around his own waist, knotted it, slipped the end of the line through the top ring on the buoy float, made another knot and then handed it off to DeCapua. DeCapua tied another float to Hanlon's waist in the same manner, and then looped a second rope around their waists as a backup. Now they were a human chain; their fates were tied.

Morley grabbed Bob Doyle by the hood.

"We stay on this thing as long as we can!" he shouted. "You got that?"

Bob Doyle nodded.

"If we have to go in the water, we go off the port bow. We go up on the railing, then we jump off."

"Okay."

The starboard railing was now entirely underwater. Morley shouted, "Let's move!" Like crabs, they clawed up the tilting deck to the gunwale on the port bow.

"Okay, listen up!" the skipper shouted. "We jump when I tell you guys to jump!" Then, to Bob Doyle: "Where's that 121?"

"Right down here!"

"Where?"

Just as Bob Doyle picked the beacon up off the deck where he had put it down for a moment, a cable snapped overhead and cracked on the deck not five feet behind him. He whirled, and as he did, a wave surged over the bow and swept the EPIRB out of his hands and clean over the gunwale.

"*Oh, shit!*"

"What?"

"I lost it! The EPIRB!"

"Get up on the gunwale!"

"*Goddammit!*"

"Get on the gunwale!"

He swung a leg over the railing, then the other.

"Bob, get on the end!"

They lined up, crouching, holding fast to the railing. The boat was keeling now at a forty-five-degree angle. Half of the deck was underwater.

"Oh my God," DeCapua was stammering, "oh my God, oh my God, oh shit, oh shit, oh shit, oh shit—"

"Cool it!" Mork screamed.

Bob Doyle looked over at the pilothouse. The emergency lights were still on. He turned and saw Hanlon, clutching the other EPIRB to his chest, his eyes shut.

"Listen!" Morley shouted, holding his hands cupped. "As soon as I say 'go,' we all go in together!"

"Oh shit, oh shit, oh—"

"When I count three, let go of the railing, take a step backward and jump! Stay calm! The suits will hold you up. Okay? Now, on the count of three, we all go in together!"

Bob Doyle looked over his shoulder. The ocean was so dark he could not tell the difference between a wave and a trough.

"Everybody ready?"

They could fall fifteen feet or a hundred.

"*One!*"

They could jump in front of a breaking wave and be smashed against the hull.

"*Two!*"

The ship was tipping, starting to roll.

"*NOW!*"

Into the abyss they leaped.

T W E N T Y - S I X

At first, all Bob Doyle felt was the cold. Cold, cold, cold like he never wanted to experience cold. It was a hurting cold, a vicious cold that had already begun deadening his toes, working its way up into the calves of his legs and setting in under his knees; a cold that ached in his ribs, that numbed his spine, that tightened on the temples of his forehead like a vise grip.

The cold squeezed and squeezed and squeezed and squeezed and all the time there was only blackness and the sound of bubbles and the *bump-bump, bump-bump* of his heart. There had been a sharp pain in his elbow immediately after they had jumped off the gunwale, but now the cold had numbed it away and there was only the tightening cold on his feet and hands and eyelids and no sound at all, only blackness and the increasing pressure and then he felt his face sharply throbbing and then the throbbing passed and it was all right. He felt wrapped in darkness, twirling, falling without end, and then he felt the kick. It was a light kick, a soft kick, and then he felt it again and he thought to kick back and he was kicked again and he thought perhaps one of the others was trying to communicate with him. Then he felt a heavy weight on his chest and it occurred to him that he might drown. I've got to get air. I've got to get to the surface, he thought, and then he realized he did not know which way was up and suddenly he was very afraid. He began kicking his legs and thrashing and fighting the water in a heavy-footed panic and he soon felt hollow and sick from the kicking and he thought: Where in God's name am I? It horrified him to think he could be swimming toward the bottom of the ocean and so he stopped kicking. The pressure on his head began to lessen, which made him wonder if he was still alive, or dying, and then he reasoned that he could not be dead yet if he was worrying about it, and then he felt pressure on his neck. Something was tugging sharply at his neck. It tugged and tugged and soon he could not fight it anymore and let himself go. In that instant he burst through the surface.

He knew it because of the noise. There was a high, moaning shriek all around, and through that noise a thundering, avalanching sound, as though several buildings were imploding around him at once. He threw his eyes open; they burned from salt. He tried to breathe; salt water flooded his mouth. He coughed, hacked, gagged.

And was under again.

Once more there was only the muffled sounds of bubbles and water being thrashed. It felt so calm and pleasant—except for the hot pain in his lungs and the pressure on his neck—and then he popped back into the world of shrieking blackness and swirling, stinging spray.

This time, he did not open his eyes. He gulped air, choked on salt water, gulped some more, coughed. All over his face he felt bites; he imagined he was looking straight into a sandstorm or standing in a field of famished locusts. Then he heard Mark Morley's voice—faint, but clear.

"*Sound off!*"

He could not utter a sound, only gulp for air.

"Hey! Sound off! Giggy? Dave?"

"Here!"

"Mike?"

"Here!"

"Gig?"

"Here!"

"Bob?"

He tried to shout but his voice sounded tiny. "*I'm here! I'm here!*"

A wave threw them together. He kept his eyes open for more than a second, and in the stabbing, blinding quick-flash of the strobe, he had seen Mark Morley's face, contorted, lips quivering, skin a bluish white. His glasses were gone; he had lost them in the jump. Without them, his eyes were wide and staring like those of a frightened child in a dark closet.

"Bob," Morley said, "how's my zipper?"

"Why?"

"Just check my zipper."

Bob Doyle took hold of him by his shoulder and patted his chest until he felt the metal tab. He could feel the skipper shaking.

"Your zipper's up."

"Shit," Morley said, "then my suit is ripped. I feel water getting in."

"Where?"

"In my legs," Morley said. "My right leg. I can feel water getting in. God, it's cold."

Just then a wave buried them. It was as though a dump truck had unloaded a ton of soaked, frigid towels and they went down so fast that a hot, white pain tore through Bob Doyle's head. He was going down, down and down and nothing but down, his legs and arms tangling in the ropes as he went, his arms flailing and his heart thumping wildly, and then he thought: Where's the rope? I don't want it around my neck. I don't want the rope around my neck.

———

Now he was sitting across from his father in the old skiff and the oars were dipping and raising silver drops from the Connecticut River. It was the day he had caught his first big perch. They drifted up to a sandbar that poked out of the river sometimes and his father took a long twig and stuck it through an empty potato chip bag and told him to plant the flag on the sandbar. He did it and together they declared the sandbar Potato Chip Island. It was their island, only theirs, and his father told him to remember it and to return to it in his mind whenever he needed time away from the world.

Later they were paddling back along the river and his father pulled in the oars, looked at him with those still, cold eyes of his and said, Bobbie, stand up, son. So he stood awkwardly in the skiff and his father looked at him darkly. So, you want to learn to swim? When he nodded his father then said, C'mere, and before he could move he felt the big, powerful hand shove him and he went in backward and sank quickly into the coolness, then cold. He fought the dark and had almost started to take a breath of water when something had him by the wrist and lifted him, dripping, into the skiff. His father was laughing. Not afraid of the water, are we? Later, sitting in the stern of the boat with the perch flat-eyed at his feet and the trees on the hills amber, he felt quite certain that he would never drown.

Spray like buckshot whipped his face. He'd come up again.

"Sound off!" It was the skipper's voice. "Bob? *Bob!*"

"Here!"

Then he felt the weight on his shoulder and turned and saw Mark Morley clinging to him.

Morley said, "When will the Coast Guard be here, Bob?"

Why do you keep asking me that? Bob Doyle thought. They probably *aren't* coming. But he told Morley, "They'll send somebody for us."

"When?"

"Within the hour."

"You said that before."

Morley's teeth were chattering. Bob Doyle could hear them. "Did I?"

"Yes, you did," Morley said.

"Well, like I said, they're probably on their way right now."

"God, I can't see anything."

"None of us can. There's nothing to see, anyways."

It was true, too. Bob Doyle could barely see his hand before his face, except when the strobe flashed. And it was flashing only every twenty seconds or so.

Gig Mork began hollering.

"Look! Look!" He pointed behind Bob Doyle and kept yelling. "There! See what I see? Look! *There she goes!*"

Between waves and flying froth and sleet they snatched glimpses of a long, thin silhouette—the last of the *La Conte*'s hull. She's a wood boat, Bob Doyle thought, and when she goes under she's going to groan something awful. But she went down without a sound. The deck lights and portals glowed as her great bulk slipped beneath the surface and hung there in the depth of the water like a huge purple bird, and then settled slowly. They all watched her go down, getting smaller and smaller, until her lights were out of sight.

The seas would not stop jumping up and down. They clawed their way up one watery hill after another. Sometimes the wave would break down on top of them. Other times they would reach the crest and then go skidding and tumbling down the back side of the swell into a cauldron of spray and foam.

I got to stay afloat, Bob Doyle kept thinking as he slugged and kicked his way through the water. I got to stay afloat. God, I wish it was daylight. But what good would light do? You can't see shit anyway with all this wind and water. And we got a light. We got the strobe. And what else do you need to see out here other than a helicopter?

Just keep your eyes closed, he told himself. As much as you can, anyway. Too much salt water in them and they'll swell up and never close.

"Bob!"

Morley had been dragging behind and swallowing water. Bob Doyle spun, grabbed him by the waist and lifted the skipper up on his chest. He put a hand over Morley's mouth to shield it from the sleet and spray.

"Breathe," he said. "That's the way. Breathe, man. Good. I'm here, Mark. I'm here."

Morley coughed and hacked.

"You all right?"

"I'm cold, man. I'm so cold. Are the Coasties coming, Bob?"

"Sure," Bob Doyle told him. "On their way."

"I'm so cold."

"How are your legs?"

"Heavy. Pulling me down. I can't hardly feel them."

So, Bob Doyle said to himself, it has already started. And how long have we been in the water? Ten minutes? He put his arm around Morley's broad back, pulled him up a bit and leaned back on an angle so that they floated together.

"Bob?"

"Yeah?"

"I hope those Coasties get here soon."

"They will."

"I hope they're coming, Bob. I'm freezing, man."

"A chopper's coming."

"I got to get on it. I got to get on that chopper."

His teeth were chattering so hard Bob Doyle wondered if they would chip.

"We'll get you on it."

"I got to get on it."

"We'll get you on it."

A wave swept over them and they lost each other. Bob Doyle counted to twenty before he popped through the foam. Morley came up beside him.

"Help!"

"I'm right here, Mark. I got you."

"Where are the guys?"

Bob Doyle pulled the skipper back up on his chest. The hood of Morley's sweatshirt was down over his face. Bob Doyle pulled it up a bit. When the strobe flashed, he saw that Mork had Hanlon by the shoulders and was supporting him the same way he was helping Morley. Hanlon was retching seawater.

He heard Mork say, "Just stay up here on my chest. I got you good."

"I can't keep my head up."

"Giggy, what's wrong with Dave?"

"His pillow didn't inflate right," Mork yelled. "He's puking all over like a son of a bitch."

"*Mike!* Can you see me?"

DeCapua yelled back: "*I can't see shit!*"

"Okay," Bob Doyle said. "We got Mike."

DeCapua's face was covered in hair. It looked as though someone had stuck a pile of seaweed to his face. All Bob Doyle could see was DeCapua's nose poking through a mane of hair.

"Hey, Mike," he said, "I like what you did with your face!"

"Wiseass!"

Morley said, "Bob?"

"Yeah, Mark?"

"Don't let me go."

"I'm not."

"I don't think I can swim anymore."

Out of nowhere a landslide of water buried them, then another, and countless others, sending them tumbling as if they were inside a washing machine. The water was very cold. They could hear the waves coming and sense when they were going to break but they never saw what hit them. As soon as they came up they took breaths and went down again. It was quiet underwater but each time they went down it was as though the ocean had peeled another layer of heat off their bodies. The ropes tying them together began to slip.

David Hanson *(left)* and Dr. Michael Propst *(right)* were instrumental in identifying a batch of human remains discovered on an island north of Kodiak in August 1998. The remains turned out to belong to the lost crewmember of the doomed fishing vessel *La Conte*. *(Photo by Todd Lewan)*

At his desk in the fingerprint lab of the Alaska state crime unit in Anchorage, Walter MacFarlane examines a photograph of a dime-size piece of fingertip skin found on Shuyak Island in August 1998. Using the image, MacFarlane and his staff identified the remains of a fisherman who died in the sinking of the *La Conte* nearly seven months earlier. *(Photo by Todd Lewan)*

Bob Doyle, a twenty-two-year veteran supply officer in the U.S. Coast Guard, joined the crew of the *La Conte* fishing vessel a year after he was forced to retire from the service. In this photograph, he is seen walking in the woods near Bellows Falls, Vermont, his hometown. *(Photo by Todd Lewan)*

In her trailer home outside of Kodiak, Alaska, Bob Doyle's ex-wife, Laurie, relaxes on her day off with their two children, Brendan *(center)* and Katie *(left)*. *(Photo by Todd Lewan)*

Mike DeCapua, his trademark mane of hair flying in the wind,
works the lines on a boat as it scuds across the waters of Sitka Sound.
DeCapua, a fugitive from the law in Idaho and Washington States, arrived in Alaska in
1983 and soon became a skilled deckhand on longline vessels. *(Courtesy of Don DeCapua)*

The *La Conte* glides past a floathouse in the calm inside waters of southeastern Alaska in April 1990. In the 1980s, the bulwarks of the seventy-seven-foot-long schooner were painted black, and four new booms and two masts were added. The extra booms made the vessel considerably more top-heavy and less steady in the rougher waters of the open Gulf of Alaska.
(Courtesy of Jeff Berg)

William "Gig" Mork, the "foreman" on the *La Conte* and the right-hand man of skipper Mark Morley, was known around Sitka as a hard drinker and as a hardworking, experienced deckhand. *(Courtesy of Edith Mork)*

Mark Morley, the *La Conte*'s skipper, kneels beside a campfire in the Alaska outback in 1997. Born in a coal-mining town in West Virginia and raised in the suburbs of Detroit, Morley had always dreamed of making a living as a hunter, trapper and wilderness tour guide in Alaska. *(Courtesy of Tamara Morley)*

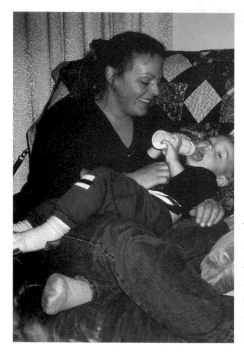

Tamara Morley bottle-feeds her infant son, Mark Jr., at her home in Sitka, Alaska, in 2000. Just before the *La Conte*'s last run to the Fairweather Grounds, she called Morley on the boat to tell him he was a father-to-be. *(Photo by Todd Lewan)*

David Hanlon, the strong, silent member of the *La Conte* crew, is seen here during a visit to his brother's home in Juneau, Alaska. Hanlon learned to fish at age thirteen from his father, who owned several commercial vessels in the town of Hoonah. *(Courtesy of Eli Hanlon)*

U.S. Coast Guard captain Ted LeFeuvre, seen here in full-dress uniform, was the commanding officer at Air Station Sitka on the night in January 1998 when the *La Conte* sank in a storm. Never more than an average pilot, LeFeuvre rose to the rank of captain through discipline and determination. *(Courtesy of Ted LeFeuvre)*

Ted LeFeuvre poses for a picture between his teenage daughter, Michelle *(left)*, and his teenage son, Cam *(right)*, outside his California home in the mid-1990s. *(Courtesy of Ted LeFeuvre)*

Russ Zullick, who copiloted the second rescue helicopter during the *La Conte* mission on January 30, 1998, shows a visitor the complex instrument panel of a Sikorsky H-60 Jayhawk, the Coast Guard's heaviest and most sophisticated helicopter. *(Photo by Todd Lewan)*

A ground-maintenance crew gives a Jayhawk helicopter a checkup inside a hangar at Air Station Sitka. The Sitka ground staff received a medal of commendation for their work as a group on the night of the *La Conte* sinking. *(Photo by Todd Lewan)*

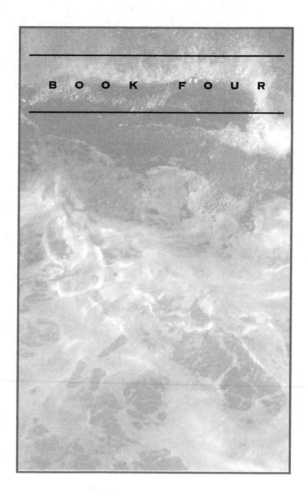

BOOK FOUR

The house sat back from the road at the end of Verstovia Street. It was a two-story Cape Cod, quite big by Sitka standards, and had gabled roofs over each of the upstairs bedrooms that gave the house a quaint, New England look. It had steel gray shingles with white trims and gutters and a white picket fence that closed in a big front yard. There was also a wrap-around porch with a spoked, white-painted banister and white posts. Window boxes hung in a row along the banister. In the spring and summer the boxes always had cheerful pansies in them and, in the winter, after a storm, the bunting of freshly fallen snow.

It was a roomy and fine house for any couple, though it could easily accommodate a family of four. There was a master bedroom upstairs and two ample bedrooms at the other end of a hall, three bathrooms, a spacious kitchen with an island counter for entertaining, an ample dining room, a utility room, a smokehouse off the two-car garage, a greenhouse out back and, his favorite room, a carpeted den with large windows decorated with lemon yellow cornices and drapes that looked out on the front porch. From the den on winter nights you could hear the wind coming in from the sound and whispering in the tops of the spruce trees outside, and if you went to the window you could see the moving shadows their branches made on the white, frosted lawn from the light of the moon.

A man named Ted LeFeuvre, who was a captain in the Coast Guard, had rented that house and lived in it since his transfer to Sitka in June 1997, the month he had taken over command of the air station. After one has lived many happy years in large, family dwellings it is not always easy to make the transition to a bachelor's apartment, and Ted LeFeuvre, who had always taken great pleasure in homemaking and entertaining company, did not want to give up having friends over for a meal or having out-of-towners stay a few nights in a spare bedroom.

Sometimes he would invite guest speakers at the air station to stay over, or the boys from high schools around Alaska who were in town for a swim meet or a basketball game, or on rare occasions even his parents, who had already made the trip once from Los Angeles. On Wednesday nights he led a Bible study group and sometimes he would cook dinner for as many as fifteen of his fellow parishioners. He liked to prepare Cajun food for his guests, seafood jambalaya, red beans and rice being his preferred dishes, and he always derived a certain pleasure when his guests commented that his cheese nachos were bordering on being a bit too spicy even for the average Mexican palate.

When there was no one to cook for he would come home from work late and fix up something easy—a can of green beans and salted broccoli, usually. As a rule he disliked microwave food—it always came out too bland and mushy—and most fast food, except for an occasional slice of pizza with vegetable toppings. Some nights he would stop at the Sea Mart deli on his way home from work and pick up one of those sterile premade salads and a few packets of dressing. He always thought that meals were, in their essence, occasions best suited for family or groups of friends. Preparing a meal in his big house merely for himself, he thought, not only was boring but also bordered on selfishness.

Having lived nearly twenty years with a woman whose hobby was painting and etching, Ted LeFeuvre took care to fill his house with paintings and charts and photographs. He liked landscapes, particularly scenes of solitary cottages, or dark forests, and on nights when the house felt a little vacant he would stand and study a larger canvas on the wall of his den, which depicted a large oak tree standing tall and strong in a hazy, Georgia field, protected by a fence of barbed wire.

He also hung a number of photographs of his son and daughter on the walls of the den and his bedroom, each showing a different stage of their development, and he dressed up end tables and counters and shelves with portraits of his parents, his brother before his skiing accident, his cousin and his cousin's wife and children. There was but one reminder of his former wife on display and you saw it in a glass frame as soon as you entered the den. It was a little gift she had given him years earlier—white, crocheted flakes of falling snow, each flake in the shape of a teardrop.

In the winter, when the northers blew and the temperatures plunged the house was kept warm and cozy by an iron stove in the den. The stove burned pellets of compressed sawdust, which looked something like rabbit food, only bigger, and it was very effective in heating both stories and much cheaper than it would have been to run the electric heat. Half the time Ted LeFeuvre would forget to refill the stove with pellets when he left for work in the morning, and he would come home to a forty-degree house. But once he got the pellet stove roaring it would take only a short time to make the entire house feel comfortable again. On especially cold nights, while waiting for the chill to leave the house, he would stand in front of the stove, listening to the norther blowing outside and watching the bright orange glow of the fat pellets burning.

Most evenings he would sit in the rocking chair beside the stove, wrapped in a comforter, reading the Book of Matthew by the lamp on the heavy wood stand. Sometimes he would look up while he was reading to hear the chime of the brass pendulum clock he loved so much or the wind as it moved under the eaves of the house and sang in the wires outside. As he sat in the chair he felt far away from the wind, although, really, the wind was whipping at all of the gables and working to get in through the cracks of the storm windows and the screen doors. In the chair he could feel the warmth of the stove and sometimes he would remember the feeling of bouncing and cradling his baby daughter in his lap all those years ago when he had stayed home to care for his wife, who'd had a particularly rough time giving birth at the hospital.

The pellet stove was a great thing in the winter and he relied on it and took good care of it through all of the coldest months. He hated the cold and the long darkness of the Alaska winter and the way memories could play tricks on you in the night, and the stove and his Bible helped him through it and kept him looking forward to the coming spring.

TWENTY-EIGHT

The first time Ted LeFeuvre's children came to stay at the house was the week he moved to Sitka to take command of the air station. It had been arranged for the two of them to meet their father in Seattle and to sail together on the ferry up the Inside Passage. His daughter, Michelle, who was eighteen, had been quick to tell him that she would only be staying a week. She had a new job, she said, and couldn't get time off. His son, Cameron, who was a year younger than Michelle, had invited a friend, Mark, along on the trip. Cam was still in high school but he was on summer break. He would be able to stay two weeks. Afterward, he would fly back to San Diego and spend the rest of the summer with his sister at their mother and stepfather's house.

He had offered to fly them all up to Sitka, but the kids had insisted on riding the ferry. *The ferry will be cool.* He reminded them it was a two-and-a-half-day boat ride. Were they sure? *Dad, we want to go on the ferry.* So he relented and bought four ferry tickets. The ship left at six-thirty in the evening. At eight-thirty they approached him in the stateroom, having not found a game room, casino, pool, bar or dance club. *How long do we have to be on this ferry?* They killed time together playing cards and telling jokes and catching up on all the insignificant news and scuttlebutt that parents find so interesting and teenagers so boring, and finally, after they'd spent more than fifty hours of seeing nothing but water and tree-studded coastline, the boat pulled into Sitka's ferry terminal a little after five in the morning. The outgoing base commander greeted them with the usual entourage of officers and had them driven to LeFeuvre's new home on Verstovia Street. To the kids, the trip had been a bore, a complete waste of time; their father wouldn't have traded it for anything.

They had a fine time that first week. It was almost as good as it had been in the old days, before all of the sadness. They went for a long hike on a mountain trail together and spent an afternoon at Whale Park. They drove the Jeep Cherokee up Harbor Mountain Road until they reached the alpine level, more than two thousand feet above the sea. Another day they spent hiking along the banks of a mountain stream. The highlight was spotting an enormous brown bear splashing around at the top of a water-fall.

Ted LeFeuvre was at the air station at six each morning and worked until midafternoon. The kids usually slept in, and he wanted to lose as lit-tle time as possible with them during the days. Sometimes he would pause at his desk to imagine the kids exploring around Sitka and visiting the har-bors and the sights, and it gave him a warm feeling to know that he would be home soon and cooking up salmon in the smoker and sharing supper with them.

The boys slept on mattresses in the bigger guest room at the end of the upstairs hall and Ted LeFeuvre's daughter slept on a bed in the smaller room, and it was a lot less lonely sleeping when he could hear the children laughing and chattering while they watched TV at night. They were a little shy at first and much neater than they were by the end of their stay, but Ted LeFeuvre found he did not mind the mess. They left their beds unmade and sometimes he would see the pillows on the floor and their sneakers and clothes and nylon bags scattered about their rooms, but it did not trouble him. He had many precise habits, as one develops after years of living alone, but he was pleasantly surprised to find that it felt good to have the kids break them up. He realized he would have his habits and routines long after the kids had gone.

Of the two children, it seemed to him that Michelle had gone the fur-thest away from him. She was tall, pretty, with long, wavy brown hair, her mother's bittersweet eyes and a wide, electric smile that brought a room to life. Michelle had known about her mother's affair before her younger brother and, as he discovered much later, several years before he himself had. She never lost an opportunity to take her mother's side in an argu-ment, and when she bridled under her father's discipline she would always make a point of saying *I can see why Mother left you*. Once the divorce was final and her mother began openly traveling to New York to see her lover, Michelle was left behind on her own for weeks at a time, a freedom her high school friends considered a real coup. Ted LeFeuvre feared his daughter would become a little too wild and urged her to live with him in Los Angeles, where he was stationed at the time, but she told him with the terrifying flatness of a fifteen-year-old: *Dad, I could never live with you because you have too many rules.* Instead, her visits grew less frequent. Under the custody agreement, he was to pay fifteen hundred dollars a month in child support and was to see Michelle every other weekend. He was careful to pay the travel expenses each time Michelle made the trip from San Diego to Los Angeles, but he saw his daughter maybe ten weekends a year, if that. She would call sometimes a day before he was expecting her, and bow out of the visit with the patent line, *Something's come up and I can't make it.*

But what really hurt was how his ex-wife had enlisted his daughter to spy on him — more precisely, on his checkbook. After one of his daughter's visits he knew to expect a phone call from his ex-wife and to be peppered with questions about his church donations. *Why do you always do that? You could give me the money.*

The boy was another matter. Cameron LeFeuvre was a copy of his father physically, reduced in scale and shortened. He was small and dark-haired, with his father's bushy, black eyebrows, the same thin, wide-set eyes and angular, unfrivolous jaw. Cam was thoughtful and he had a sense of loyalty and was a quiet, respectful boy, although in repose his face had a mischievous look to it. He compartmentalized his feelings and, like his peers, had a defiant streak and tuned out talk that did not focus on cars, the beach or girls. After the divorce he moved with his father to a three-bedroom duplex in Redondo Beach outside of Los Angeles and they lived together until word came from command that Ted LeFeuvre was going to Sitka.

The night before his ex-wife and her new husband were to take Cam back to San Diego, Ted LeFeuvre prepared a dinner of steamed broccoli and red beans and rice and chicken nuggets. Cam ate in silence. Ted LeFeuvre sipped his Dr Pepper and watched him, searching for something to say.

All he could think of, finally, was, "So, there's nothing I can say to get you to change your mind?"

The boy shook his head.

"You could bring Buck." That was his son's yellow Labrador, which he had raised from a pup.

"No, Dad. No. I couldn't."

"You know, I always prayed that your mother would come back."

"Yeah?"

"We can't always get what we pray for, I suppose."

There was a heavy pause, and then Ted LeFeuvre said, "Sure you don't want to go to a godforsaken hole?"

His son laughed.

"You should never be afraid to change your mind, Cam. Especially when it comes to moving to a cold, dark, rainy, bear-infested—"

"Dad."

"Yeah?"

"I'll miss you, too."

In that way he had lost his only son. A decision was a decision and a compromise was truly a compromise, and knowing all of this and having been taught all of the other pleasantries of divorce, Ted LeFeuvre was happy that at least the children could come to see the change-of-command ceremony in Sitka. If one week with your daughter and two weeks with your son is what you get, he thought, then that's what you draw. Besides, a lot can be shared in fourteen days with the people whom you cherish above anyone else in this world. But why split up in the first place? You had better leave that one alone, he told himself. Be glad that you have two healthy children who know you and respect you and show some gratitude to the Lord that you were able to raise them for seventeen years, even if the last three brought you so much pain. They are fine kids, on balance, and they got some of her good qualities, along with some of her bad ones, but that is all part of it and nobody is so complicated that you cannot at least try to understand what makes them tick.

He had always been able to view the world through rose-colored glasses and he normally coped with change and hardship by looking ahead. He was one who got more enjoyment out of planning a trip, looking ahead to the trip, than the actual trip. Perhaps that was a character flaw, a mechanism for self-preservation. But when it came to his children or his ex-wife, looking ahead did not seem to free him of the regret or the loneliness. He had learned to stop the guilt for missing the signs of trouble in his marriage, but there seemed no end to the regret and the loneliness, not even through prayer.

Whenever his children were with him was like taking a vacation from the loneliness, or rather, the aloneness. He knew that each time the children came to visit they would open up new rooms in his heart for the aloneness to move into once they were gone. But there was nothing to do about that. It would be very empty and bad for a while, but no good would come from fearing it.

Michelle caught her flight out to California a day after the ceremony. After she left it rained almost every day for a week. The mountains were misty and the wind raged out over the sound. Ted LeFeuvre worked hard and Cam and his friend took walks along Halibut Point Road to the ferry terminal and fished the bay there. One day Wayne Buchanan, the executive officer at the air station, took the boys out for a full day on his boat and showed them a spot he knew where the reds ran thick. They caught many salmon and cooked them in the smoker at the house.

Then one morning Ted LeFeuvre put on his blues and packed the boys' bags in the Jeep and drove them to the airport. The boys checked in and stepped through the metal detector. Cam turned, waved and mouthed, "Bye, Dad!" and then walked through the doors and into the boarding lounge.

So they were gone, too, and Ted LeFeuvre eased the Jeep out of the parking lot and rode next door to the air station. Wayne Buchanan knocked on his office door.

"The boys leave yet?"

"I just saw them off."

"That Cam's a good kid. Too bad he didn't want to go to school up here."

Ted LeFeuvre looked out the window.

"I guess you're really going to miss them."

"Yes," Ted LeFeuvre said. "I guess I will."

TWENTY-NINE

His first year in Sitka, Ted LeFeuvre worked as hard as he ever had in the Coast Guard. It was not a large air station, as bases went. There were three hangars, a trio of helicopters, a runway, three landing pads, an operations center and, down the road, the gymnasium, officers' club and barracks. But he had 150 men and women to manage, and Sitka was not easy on outsiders. One could get claustrophobic, with the rain, long winter nights and the creeping detachment that came with living in the middle of an archipelago with no road network, only water and more water. Even locals left the island in the winter for a couple of weeks in drier, sunnier Seattle—*Seattle*, of all places. Getting off the Rock, they called it.

Truly, the country could be harsh on the soul and, knowing quite well how easily one could feel cut off from the world, Ted LeFeuvre took extra care to engage his staff consistently. He made a point of visiting the pits and machine shop at least once a week, not a common practice among his predecessors, and he enjoyed chatting with machinists, mechanics, janitors and the cooks. He routinely ate at the enlisted mess—he rather liked the food—and around holidays he hosted parties, albeit alcohol-free ones, for his officers. He took care not to go off on tangents, lock his door or spend all day on the computer. And yet he did not coddle his people or make excuses for them; he was the commanding officer, after all, and he could not allow himself to get chummy with subordinates.

His pilots, to his good fortune, he considered to be tops in the business. He had fifteen fliers, transplants from the navy, army, Marine Corps, and the usual recruits from the Coast Guard's farm system. All of them had completed two tours before Alaska and not one was a hotshot. Usually you would get at least one air cowboy with a cute temper. But in Sitka he'd been blessed with men who flew with their heads—not their balls. He had no patience for tricks or stunts or those who played up the risk of their missions in their flight reports. Everyone knew what it took to climb into a helicopter on a bad night in Alaska; there was no need for grandstanding.

He had never particularly liked accepting awards for doing his job — it had made him awkward, in fact — and he liked even less having to rank pilots when it came time for their evaluations. There were a handful for whom he held a special regard, but he never let them or anyone else know it. One was Jack Newby. Ex-army, tall, with the composed expression of a ceramic cat, Newby had a wit that could make skulls smile. He also had more confidence than anybody Ted LeFeuvre knew. Newby was not concerned about leading, yet he was the kind of pilot who could get his crew to follow him anywhere and do anything. What he knew about helicopters he knew absolutely and what he did not know he was unashamed to admit. Oftentimes there was something hollow about charismatic people. But in Jack Newby's charisma there was nothing but sincerity.

Another was his deputy operations officer, David "Bull" Durham. Bull never said much but he was perhaps the hardest-working officer Ted LeFeuvre had ever met. He was conservative in the cockpit, flew strictly by the book, and yet was more technically resourceful than the daring pilots. He was a human reference book in the cockpit. He knew precisely how far the aircraft could be pushed under any condition and his ability to use the whole helicopter, to maximize the machine's output, was unmatched. He did not have the natural charisma that Jack Newby did. But Bull had a huge heart and perhaps was the more knowledgeable.

One of his very best aviators, Dan Molthen, combined superior flying instincts with a refreshing lack of ego. Those instincts, Ted LeFeuvre thought, must have come from his days in the navy, where he did a lot of night flying off aircraft carriers. At thirty-five he won the Distinguished Flying Cross, the highest aviation honor in peacetime. He was lean but strong — he worked out like a fiend — and had a long face that made an angular, sober silhouette. He was boyish, though, so self-effacing that unless you flew with the man you would never know he possessed motor skills that most aviators could only dream about.

Then there was Bill Adickes. Crack was a short man, fatless, with smoldering topaz eyes set in a near-permanent squint. He didn't have outstanding motor skills like the others, and he wasn't as knowledgeable on the technical aspects of the aircraft. What he had was imagination and cunning. He was also economical with his helicopter, something he'd learned by flying combat helicopters for twelve years in the Marine Corps.

Sometimes Adickes got himself into trouble for not planning all of his moves out in advance. But he was so unflappable, so creative, and so very, very lucky when he needed to be that he always—always—pulled himself and his crew out of jams to complete a mission.

As for the rest of his fliers, the one Ted LeFeuvre thought might be closest to entering that top tier was a thirty-three-year-old from Newton, Massachusetts, Steve Torpey. His peers called him Torp, the made-for-TV pilot. He had a wide, telegenic smile, and some remarked that he'd come up to Alaska to forge a reputation as a top gun. Ted LeFeuvre hadn't seen any of that in him. He was a little stubborn, perhaps, prone to impatience, maybe, but he was diligent and knowledgeable. And if there was such a thing as a natural, Torpey was it. He had extraordinary intuition in the air and fast, fluid motor skills. In four years in Alaska he hadn't gotten a single big case, though, and it bothered him. But once he gets that case, Ted LeFeuvre thought, he will have something inside of himself for the rest of his days and it will make everything else he does in life easier.

All of his pilots, of course, had come to Alaska specially trained to fly the Coast Guard's most sophisticated helicopter, the Sikorsky H-60 Jayhawk. It was a tank of an aircraft—it weighed twenty-one tons without a crew or gas—and had the power of one. With twin, 1,980-horsepower engines, the Jayhawk cruised at 146 knots and hit dash speeds of 180. It could fly 325 miles offshore, hover forty-five minutes and still return with a safe fuel reserve. Its computer could fly and land the aircraft, its global positioning system could receive data simultaneously from four satellites and its forward-looking infrared sensors made navigation a snap.

The Jayhawk certainly was to his liking. Flying it required a certain modesty at the controls, a careful touch—not brute strength. Though faster than its predecessor, the hulky H-3 Pelican, the Jayhawk was a wide aircraft susceptible to downdrafts. Once in the water, the H-60 sank like a stone. You either got out of the aircraft in thirty seconds or you went for a ride to bottom of the ocean.

Ted LeFeuvre had never ditched a helicopter. But he was never more than an ordinary pilot and not much to talk about at sea, either. He got violently ill his first cruise aboard a Coast Guard cutter. He threw up halfway through his inaugural flight in a T-34 Bravo helicopter, and left his dinner in his lap during a night flight in a T-28 plane. (The engine caught fire and he was greeted on the runway by a squadron of fire trucks.) At flight school in Pensacola he was flunked for bringing his T-57 Jet Ranger in too fast on an approach and had to fly extra time to compensate for his mishap. He barely graduated with the class of 1977, his grades one one-hundredth of a point above passing.

He'd always taken longer to learn maneuvers that other pilots picked up effortlessly. While training in New Orleans to qualify to fly the H-65 helicopter, he was sent home after a few blunders and grounded for two days. For years he cringed at hovering close to the water; he still tightened up a little when flying in tight formation with other aircraft.

Physically, he was no recruiting-poster type. He was unathletic, not particularly tall, wore glasses and had gone gray. At forty-six, he was in okay but not spectacular shape, though of late he had begun to feel a slowness, a fuzziness in the perimeters of his reflexes. He chided himself for not getting more time in the helicopter, but it didn't bother him too much, really. He was an administrator now more than a pilot. It would be silly for him to stand in for his aces.

T H I R T Y

Most nights he had trouble sleeping. When Ted LeFeuvre had been married he would go to bed early to read a book and to be together with his wife. She had wonderfully beautiful hair and a lovely face and body and he would lie sometimes and watch her hair shine as she twisted it up in the lamplight, and if he could feel the warmth from her arms and legs he would be happy. Besides the wildness there were all the small ways of making love and he always looked forward to lying next to her so that he could revisit them. But now he no longer read in bed and he would stay up late to put off lying in bed alone.

Often he sat up in the den. Some people said keeping a journal might help him get over things. Rubbish, he thought. It had been bad enough losing his wife. Why would he want to record it now? So he read the Scriptures and the Psalms. There was one in particular, Psalm 121:1–2, he found useful: *I lift my eyes up to the hills. Where does my help come from? My help comes from the Lord, the maker of heaven and Earth.* Other tools that helped were his John MacArthur tapes and the R. C. Sproul book *Willing to Believe.* He would rock in his chair, listening to sermons and reading, and ponder questions that had long remained unresolved in his mind.

God, he thought, how much of my destiny is of my own choosing? Do I have the complete freedom to choose my destiny? Did my ex-wife? Or is it only You, Lord, who decides our fate?

They were vexing, sobering questions for a man of the military who was only then realizing the enormous task of changing one's course in life, and they prompted, to his chagrin, other troubling thoughts. If all men, as he had always believed, were inherently imperfect beings, what gave him the right to expect his former wife to be a perfect, loving spouse for always?

Night after night he tussled with this moral dilemma. What did it really mean to forgive? What did forgiveness look like? Perhaps he was too logical and analytical a man to forgive. He wanted to forgive his ex-wife. But if he did, would that mean he'd have to be her best friend? Call her all the time? Commiserate with her? Shake hands with her new husband? He didn't want any of that. He was through being a sap. And he genuinely disliked the man his children referred to as Jughead. But if he felt this way, did it mean he had not truly forgiven?

He tried not to anguish over it, and as the months passed he gradually ceased to anguish and did his best to free himself of guilt through work, insofar as he could. In fact, he found he was able to replace nearly everything but the children with religious discipline and hard work. When he was lonesome for the kids he would pray for them and remind himself to stay within the carapace of work and prayer that he had built for his own protection. Work hard, he told himself when he felt the loneliness coming. If you stop working hard now you may lose it. Most of the time the carapace held up fine. But there were slow periods at the air base and holes in his schedule between church events and meetings, and it was during those periods that the aloneness was worst.

He realized that it was probably normal to feel lonesomeness for his children as he did, being so far away from them for so long. He was unafraid of dying alone, or even dying in disgrace. But he was terrified of the destructive nature of loneliness. It opened the door to self-pity, which he knew always led to other moral weaknesses, and they in turn would cause him to fail the Lord. Not that he considered himself on the verge of a moral meltdown. But he was wary of one.

Some nights when sleep was elusive he would go to the computer he had set up in the spare bedroom and write to his parents. His father's heart was failing and the doctor had said he could go at any time. It bothered Ted LeFeuvre that he could not be there for his father in California at that moment. When he began his letters he would feel connected to them, but as he signed off the joy would drain out of him the way the tide slips off a flat and the ebb begins in a channel.

Above the computer, on the shelf, he kept a book, *The General Next to God: The Story of William Booth and the Salvation Army*. Sometimes he would flip through the pages, reading about William Bramble Booth, the pawnbroker's apprentice who began the Army's work in the slums of London in 1865 at the age of thirty-six. A handwritten note on the first page always made his heart ache: *To dearest Ted—Hoping you will get some further inspiration from the life of your great-great grandfather. He would be proud of your dedication to the Lord. Mother and Dad, Christmas, 1972.*

Whenever he read these words, he thought: My dad is dying, my only brother, Phillip, had a severe stroke after a skiing accident, and I'm the one crying like a baby over a divorce. Lord, my ancestors would be ashamed of me. His great-uncle had served as William Booth's chief of staff at the Salvation Army. His great-aunt was once the U.S. commander of the Salvation Army. His great-grandmother Catherine Booth had worked at her father's organization in London until she sailed to America to start a Bible college in Kentucky. His grandparents were missionaries in America, Palestine and in the Upper Volta, where his grandfather died of dysentery and malaria in 1924. His uncle had been bishop of Manchester in the Church of England.

He had planned to enter the ministry himself after graduating from Azusa Pacific, a Christian college, where he'd been a youth pastor and had double-majored in psychology and biblical literature. But at the last moment he decided against it. He wasn't exactly sure why at the time. It suddenly didn't *feel* right—a shift in attitude he chalked up to divine intervention. It was only years later, after losing his wife and kids, that he figured out that his inability to respond to others' emotional troubles in an empathetic way was what had really kept him from becoming a pastor.

It was painful to admit this, even more so than going through life with a reputation for being a logical, analytical kind of man. He was actually a man of deep feeling. But he had always been a man of very strong will, a man always in complete control of his emotional responses. So strong was his character that over time he had grown unaware, in a boyish sort of way, of other people's emotional traumas. How another person could slip so far out of control was a foreign concept to him. It simply didn't register.

Not even when it happened to him.

The lady who led him to the brink he met, not surprisingly, at church. Her name was Katherine Ballish. She'd moved to California from Oregon as a fourteen-year-old, and he often saw her and her older sister at Sunday services. She liked to be amused and she had a deep, uninhibited laugh and a blitheness he found peculiar. They saw each other at Bible study and he frequently ran into her in the halls of Le Habre High. In his senior year he had the minimal wit to surmise this was no accident; when he asked her about it she told him she'd leave early from her classes and run to catch him as he left his. Even so, he remained oblivious to her sexual interest until a buddy of his pulled him aside a day before graduation and told him that Kathy had the hots for him.

During his junior year in college they moved into a tiny house next door to East Whittier Friends, the church where they'd married. She found work across the street as a secretary for the denomination's supervisor. He sent résumés to seventy-two schools up and down the California coast looking for a job teaching and got no bites. (He couldn't speak Spanish, so he was considered unqualified.) So he took a temp job for the Western Auto Supply Company, read gas meters for the state utility and delivered concrete block to building sites before finally filling out an application to the Coast Guard in November of 1974. Six months later, at his wife's insistence, he called the recruiting office. "Sorry for not calling you back, Mr. LeFeuvre," said the petty officer who answered. "We lost your phone number. You've been selected."

So began their lives as a Coast Guard couple. Kathy gave birth to Michelle weeks after their arrival in Eureka, California, his first assignment, and Cam came along a year later. After that, the LeFeuvres crisscrossed the country, moving from air base to air base. He flew rescue missions. She took up watercolor painting and crochet.

Only once, during seventeen years and nineteen address changes, did he have even the slightest clue that his wife was miserable. That was in New Orleans. One night he'd been playing a board game with the kids. She was doing the laundry. Offhandedly, he asked her why she didn't wash the clothes during the day and she flew into a rage. *My time is mine. You will not dictate to me.* And she stormed out. He thought about her outburst for a moment, shrugged it off and went back to his Parcheesi game.

In 1988 he was reassigned to Kodiak. That first year in Alaska he was away 180 days on fishery patrols. He wasn't home much the following year either. Then, one May afternoon, he came home after a monthlong deployment on the Bering and she gave him an especially warm greeting. She even invited him to dinner, to a Mexican place he liked.

They had ordered and were picking at an appetizer of spicy cheese nachos when she put down her napkin, gave him a sidelong look and said firmly:

"Jim Rao asked me to leave with him."

Rao was a pilot who lived downstairs with his wife, Leigh-Anne. In a month they were supposed to be leaving for Florida. Ted LeFeuvre lifted a nacho out of the basket.

"Well," he said, "that must have been funny." He snapped the dangling cheese with his finger and lowered the nacho into his mouth. "So what did you tell him?"

"I told him I wanted to go with him."

He stopped chewing.

"Oh."

She looked down at her plate. He finished chewing the nacho and tried to swallow. His throat was dry.

"So"—he cleared his throat—"what are you going to do?"

"I don't know."

"Well, that's a relief."

"Stop it."

He wanted to say something cute, something smart, but all that came out was, "Are you going, then?"

"I'm afraid to."

"Why?"

"I'm afraid you'll get the kids."

Years later he thought it might not have been as bad if she had not caught him in the parking lot and told him to go right back inside and pay the bill because she had no money. He also thought it might not have been as bad if she hadn't forbade him to tell anyone about her affair in order to protect her lover's Coast Guard career. It certainly wouldn't have been as bad if he had simply cut his losses right then, if they'd split and gone their separate ways.

But he couldn't. Lord knows he did not want to love his wife. But Lord knows he did and time and again he would plead with her to save their family. Sometimes she softened and promised never to see her lover again. And then the following morning he would come down to breakfast and she'd gaze blankly at him, as though she didn't know him from the queen of Siam.

He remembered in San Diego, where they transferred him a year later, how he'd lie in bed beside her and stare up at the bedroom ceiling. It was a vaulted ceiling with joists that came together at one point over their bed and he would stare at that one particular spot for hours and hours. It became his symbol of agony. The light of the streetlamps would come through the big window and illuminate the ceiling and he would look up at the way the ceiling gathered above him and the way the shadows of the leaves of the tree outside moved on the ceiling and in this see the loss of everything. He would see the loss of his family, of his children, of the complete control he'd once had over his emotions, and he would go limp and cry easily about words she'd thrown at him, like daggers.

"Have you been in contact with him again?"

"Think what you want."

"I need you to tell me you weren't with him."

"I can't do that."

It went on like that for three years. Oddly, just before the very end, it seemed things were improving. She was less preoccupied, her kisses less indifferent, and she talked about their going away someplace, even retiring together. It was during the spring of 1992, and he had to go to Alabama for flight simulations. She went to her mother's in Zebuline, south of Atlanta, and surprised him by showing up in Mobile a day before his training ended. She insisted on making love in his barracks room. He gave in, knowing that it violated regulations. The next afternoon they drove back to Zebuline without a word.

That night he lay in bed, his mind numbed from dread. The marriage was over. The lovemaking at the barracks was the coup de grâce. He knew it. He felt the way her skin prickled when he touched her. Abruptly she got out of bed and went to the living room. He followed. She was crying. When he reached out to dry her cheeks she pulled away.

"What is it?" he asked. "What do you want?"

"I want to leave you."

"To be with him?"

"That's none of your business," she said. "I don't love you. I've missed that for such a long time."

"Kathy," he said.

"No, no, no," she said. She shook her head. "I want a divorce. I want to be with Jim. I haven't loved you for the longest time. I want us to separate. I want us to have two apartments. Maybe we could live in the same apartment complex together. Michelle can live with me, and Cam with you. Cam could come over sometimes for dinner."

"Don't be ridiculous."

And then the rage flew out of her. Afterward, he lay awake. He was tired but strung out. When a faint glow lit the edges of the window shade he sat up in bed. He had it all planned out. It was a flawless plan. He'd go jogging. He always went jogging in the morning. He stood up and slipped on a T-shirt, white socks, gray shorts. She'd never suspect a thing. He tied his running shoes. He'd jog down to the closest thoroughfare. There was one not too far. He stood up and went to the door of the bedroom and turned the knob. All he had to do was jog against the flow of traffic. When he saw a truck coming at a good clip he'd throw himself in front of it . . .

"Ted?"

He turned slowly around. Kathy was sitting up in bed, peering at him.
"Come back to bed, honey," she said.

"No."

"Please, Ted," she said soothingly. "Please come back to bed."

He wanted to believe. He wanted so much to believe.

"I have to go."

"No, Ted. I've changed my mind. I have. I really do love you. I do.
Please, honey. Please come back to bed."

He hesitated.

"Please?"

Sixteen months later the divorce was official. When his promotion to Sitka
came through he was so dead to himself that he thought Alaska would be
perfect for him. The kids would be a big loss. But he needed to be away
from her. What he needed was his job, his base, his habits. Yes, his habits.
Keep your habits, he would repeat to himself as he lay in bed waiting for
sleep to take him, because you are going to need them.

T H I R T Y - O N E

On his fourth afternoon as commander of the air station Ted LeFeuvre
was at his desk poring over a mound of papers when he heard a knock at
his door.

A voice said, meekly, "Captain LeFeuvre?"

He looked up. In the hallway stood a tall, thin fellow, balding on top,
with a uniform that could have used an ironing. He had a grim smile and
sad eyes. Saddest eyes he had ever seen on a man.

"Yes?"

"Sir," the man in the doorway said, "I'm Bob Doyle, your chief warrant
officer in supply."

"Yes?"

"We had an appointment?"

"Oh, yes," Ted LeFeuvre said. "So we did. Come in. Come in and shut the door."

"Yes, sir."

The door closed with the click of a breaking icicle. Ted LeFeuvre made one last notation on the contract he was reviewing, capped his pen and dropped it in a cup. He leaned back in the chair and studied the man more closely.

"Have a seat."

"Thank you, sir."

In the chair the man lost some of his awkwardness. He did not look in awfully good shape, his cheeks sallow and right on the edge of drawn, the eyes thinly glazed with deep, purple-black smudges under them. A vein throbbed in his throat.

"So," Ted LeFeuvre said, "I'm told you've been missing work lately."

"Yes, sir."

"How many days did you miss this week?"

"Three."

"And last week?"

"Four."

He had heard things about this officer, none of them very complimentary. Nobody could keep track of him. His supervisor, Bill Adickes, an officer Ted LeFeuvre had a good impression of, had described the man as a classic drunk and, like all alcoholics, worthless.

Still, this man had made an appointment and wanted to talk. The fair thing to do was to hear him out.

"Do you mind, Mr. Doyle, if I ask why you've not been coming to work?"

"No, sir." A slow flush crept up the man's throat. "You see . . . I, uh, can't . . ."

"You can't?"

"No, sir, I—"

"Mind telling me why?"

"It's stressful, sir. It's just—it's stressful."

"Stressful?"

"Yes, sir."

"And why is that?"

"On account of my wife."

"Your wife?"

"She's been shacking up with a guy here at the station." The man let out a thin sigh and looked down at the floor. "An enlisted man, sir. A mechanic."

"What's his name?"

"Koval. Rick Koval." He hadn't taken his eyes off the floor. "Do you know him?"

Ted LeFeuvre recalled hearing the name once. "What makes you think this affair is going on? Have you ever caught them together?"

"No, sir."

"Never?"

"Not yet."

Ted LeFeuvre sat back.

"I just can't come in and see him here," the officer, Bob Doyle, said. And then he muttered, "It's awful."

"Is it?"

And then he launched into his tale of woe; how he missed his wife and how awful that was, and how it distressed him to come to work and see the man who had ruined his family, and how Koval would goad him and try to pick fights, and how Koval's friends in the hangar would ridicule him, and how much he missed his kids, and on and on, until Ted LeFeuvre looked at his watch and saw that nearly half an hour had passed. He also noticed that a sweat had broken out across the man's long, perfect Irish upper lip, and he considered it and wondered if the man wasn't exaggerating or inventing the whole thing. Ted LeFeuvre didn't smell any alcohol on him. Just the odor of stale cigarettes.

"Mr. Doyle," Ted LeFeuvre said finally, interrupting him. "You don't mind if I call you Bob, do you?"

"No, sir."

"Good."

The man's face brightened at the gentle tone of his captain's voice.

"You know, Bob," Ted LeFeuvre said gently, "you and I are sort of like one another."

"Sir?"

"Well, think about it. We're both grown men. We've both made a career in the Coast Guard. We've both had families. And we've both lost them."

"Sir?"

"You see, Bob," Ted LeFeuvre said, "my wife left me, too. And yes, she was unfaithful, just like your wife."

The man nodded solemnly. He certainly knows how to play parts, Ted LeFeuvre thought. Alcoholics are great showmen. Always good at showing people what they want to see. He continued in the same gentle tone.

"It went on for three years, Bob. Three years. Can you imagine that?" Ted LeFeuvre leaned forward over his papers. He did not smile now.

"And she took my kids." His eyes hardened. "But you know something, Bob? In the midst of all of it, I went to work."

Bob Doyle said nothing.

"Do you understand what I'm saying, Bob?"

"Captain?"

"You will come to work."

"Yes, sir."

"And you will do your assignments."

"Yes, sir."

Now the man's eyes grew furtive and the skin seemed to tighten over his face until all the lines and wrinkles and bags under his blue eyes were gone. It drew tighter and tighter as if being stretched across a skull. His lips drew in tight and what little color had been in his face drained out of it so that now Bob Doyle's skin was the color of used candle wax.

"Now," Ted LeFeuvre said, leaning back in his chair again, "is there anything else?" He glanced at his watch. He had a meeting in the hangar deck in five minutes.

Bob Doyle stared at him with those sad, remote eyes. The eyes were the only things that had not changed during the conversation. The rest of his face had fallen apart.

"No, sir," he said.

One morning that August, Ted LeFeuvre sent two of his lieutenants, Bill Adickes and Karl Frey, around to Bob Doyle's house just before lunchtime. He had not shown up for work in a week. He had not called either.

They found the front door open and the hall lights on. In the den, beside an empty bottle of Crown Royal and a pile of cat dung, they found Bob Doyle sprawled on the floor in full dress uniform, snoring.

Two months later, the police pulled him over on Sawmill Creek Road and charged him with his second DUI violation. Six months after that, Ted LeFeuvre had Bob Doyle retired.

He hadn't seen Bob Doyle since. The man had left no forwarding address, no telephone number, and had stopped showing up at the Eagle's Nest. He didn't come by the base ever again, not even to collect his retirement ID card or to sign his pension forms.

He was gone, in fact, the day they repossessed his house. All that remained were six hungry cats, animal excrement in the most unimaginable places and pizza boxes, beer bottles and empty whiskey flasks. The only furniture left was a card table and four stools—all apparently pinched from the Eagle's Nest.

It took the Coast Guard more than a year to renovate the place. Nothing inside was salvageable. Ted LeFeuvre ordered the cost of the renovation paid for out of Bob Doyle's retirement check. He also had accounting deduct an additional $125.13 to cover three checks Bob Doyle had bounced at the Eagle's Nest, the officers' club he had been in charge of managing.

That summer, Ted LeFeuvre wrote in Bob Doyle's twenty-third and final evaluation:

> *I have never served with an officer who accomplished so little and yet created so much work for others. Baby-sitting this CWO3 demanded more energy than any unit should have to endure. A driving-under-the-influence conviction, dozens of unexplained absences, and a total void in devotion to duty reflected the poorest judgment imaginable, and a completely unsatisfactory professional example for peers, subordinates and the community.*

Ted LeFeuvre couldn't remember ever having written such terrible things about anyone in seventeen years in the Coast Guard. It was unfortunate, he thought. There was no enmity at all in the man. Bob Doyle wasn't mean at heart. He wasn't combative or angry. He didn't have an ax to grind. He'd probably done enough good things to know what path in life he wanted to be on. He was simply incapable at that moment of getting on that path. Perhaps you're being too coldhearted with him, Ted LeFeuvre said to himself. No. You took the only course available. You had to put him out. It was too bad. But why didn't Bob Doyle just pull himself together? Why didn't he just come to work? Work has always helped you, he said to himself. It's the one thing you can always count on. There was no excuse for Doyle's behavior. Still, it was a shame watching the man go to pieces like that. Well, it was out of his hands now. There was nothing he could do for Bob Doyle. Not anymore.

T H I R T Y - T W O

For breakfast that morning, the thirtieth of January, Ted LeFeuvre had a half bagel with cream cheese, untoasted, and a half can of Dr Pepper. Afterward, he made his bed, folded his pajamas, tucked them under his pillow and took a shower, scrubbing his graying hair with shampoo and rinsing under the prickling drive of the sharp-jetted shower. He brushed his teeth, scraped the stubble off his jaw and chin, parted his hair neatly to one side and surveyed the face in the mirror. It was not a bad face. It was a handsome face. Well, at least he could say it was a face without pity. That much he could say.

He put on his uniform, sat at the foot of the bed and tied his shoelaces, flattened out the wrinkles he had just left on the quilt and glanced at the clock. It was six twenty-six. He went downstairs and checked the pellet stove. Low again. I'm not coming home again to an icebox, he thought. He went out to the garage, filled a used coffee can with pellets and returned to the den and poured the pellets into the stove. Then he picked a coat out of the closet, brushed a few pieces of lint from it, zipped it up and went outside and got into his Jeep. It was cold and the motor would not turn over right away.

The streets gleamed under the streetlamps. There was a heavy mist over Swan Lake. The lake is frozen, he thought, but not yet thick enough for the skaters. He made a right on Harbor Mountain Road. Rain ticked across the windshield. It was hard focusing with the rain and darkness. He was supposed to meet with Paul McCarthy, the pastor, and the other members of the church board at seven. He glanced at the dash clock. Six forty-seven. He sped up a little. He did not like tardiness or keeping people waiting. And he had to be at the air station by eight for the morning briefing.

As it was, the eight o'clock briefing turned out to be totally unremarkable. Engineering had little to report; all three of the Jayhawks had been serviced and were up and running. The duty roster had been posted for February. Staffing had been cut to half of the normal complement for the weekend. Half of his pilots were unavailable. The operations officer, Doug Taylor, was away on vacation, as was Wayne Buchanan, the XO. Jack Newby had the day off. Four other fliers had taken postholiday leave. That left seven pilots, and him. January was not a busy fishing month. Seven was plenty. A low-pressure system was approaching from the Aleutians and was expected to move through the area in the next forty-eight hours. The forecast called for *snain*—snow and rain, mixed—and twenty-knot winds out of the southwest. Typical.

Upstairs in his office he got off a good morning's work, finishing two personnel evaluations and writing thank-you letters to several volunteers he could not remember meeting. He took a call from a city official in charge of the airport expansion. The city wanted a slice of Coast Guard property so the access road could be moved and the airport building lengthened. At eleven-thirty he changed into his sweatpants and hooded jacket and sneakers and went for a run. On Fridays he jogged along the access road, crossed the O'Connell Bridge, turned around at the intersection of Lincoln and Harbor and jogged back. That day the wind was blowing just enough to annoy him. And it was sleeting. The sleet felt like pinpricks on his face.

Crossing the bridge, he glanced out at the sound. It looked gray, flat. He saw no boats coming or going. It felt good to run on his lunch break. His cheeks smarted from the sleet but he liked the burn in his legs.

After lunch and a hot shower he returned to his paperwork. They were going to rehabilitate the galley, the dining facilities, the medical area and the barracks. There wasn't enough housing for single people. He had to study the engineering plans. He took a break and went down to the machine shop, then the hangar, and then the ops center to chat with the duty officer. At two-thirty he returned to his desk and finished two more evaluations. He read over what he had typed, corrected a few things, printed out copies, signed them and put them in his drawer. He looked at his watch. It was nearly four. Another day without flying, he thought. He'd not been up in a helicopter since January 12. That was eighteen days. The last time he'd sat in an H-60 he and Dan Molthen had done a two-hour fishery patrol together. Kind of snotty out today, anyway, he thought. Try to go up next week.

He put on his coat and locked his door. He saw Dave Durham in the hallway. Bull was big. Perhaps the shaved head made him look taller than he was. Perhaps that was why they called him Bull. He had never asked him. Durham was on his way home, judging by his gait.

"Captain," Durham said, "today is Friday, right?"

"Last I checked."

"You're home tonight if anything comes up?"

"No," Ted LeFeuvre told him. "I'll be at the high school. Tonight's the big basketball game."

"Right. I forgot."

"You can reach me on the beeper."

"They're playing against Juneau, right?"

"That's right."

"Juneau's got that big kid. What's his name?"

"Don't ask me," Ted LeFeuvre said. "All I know is that the NBA is scouting him and the place is going to be packed. Police are worried about scalpers."

They laughed.

"What time's the game start?"

"Seven."

He drove straight home in the dark. He needed to get a move on. He had told Betty Jo he would stop by her house before six, and he didn't like keeping people waiting.

Betty Jo Johns was a Tlingit woman who sat next to him in church on Sundays. She'd been married twice, the last time to a Native fisherman who played around on her, and she had two sons. Some people gossiped that he and Betty Jo were an item. He enjoyed her company. He thought Betty Jo had a tremendous sense of fair play and ethics, and she loved the Lord. But he didn't want a partner, not just then. He did not miss sex yet. When would he miss it? He did not know. No one knew. That was the funny thing about it.

He arrived at Betty Jo's a few minutes after six. Her son, Jeff, was still upstairs putting on his uniform. There was a large crowd at the high school. They squeezed into a space two thirds of the way up the bleachers, not far from half court. He and Betty Jo sat and watched the boys warm up. The crowd was excited. Ted LeFeuvre sat down and relaxed. He was not much of a basketball enthusiast. It did feel good, though, to be part of a crowd. It was nice not to think on a Friday night. Thinking did you no good sometimes.

Tip-off was at seven sharp. Exactly seven minutes before the referee blew the starting whistle, a satellite 650 miles above the earth picked up an emergency signal from a radio beacon in the Gulf of Alaska.

SARSAT-4 was one of seven low-earth-orbiting satellites circling the globe on January 30, 1998. It was capable of reading analog signals from 121.5-megahertz EPIRBs, like the one triggered first aboard the *La Conte*, and digital transmissions from 406-megahertz models, like the one Bob Doyle had triggered just before abandoning ship.

That night, however, SARSAT-4 picked up only one signal from the *La Conte*—the one coming from the stronger 406 beacon.

Instantly, the satellite relayed the Mayday to a powerful computer inside the U.S. Mission Control Center in Suitland, Maryland, and the mainframe began checking thousands of boat registries to identify the ship in distress. When a beacon is registered with the Coast Guard, it takes less than a minute to determine the ship's name, its characteristics, the name of its owner, the owner's telephone number, the boat's radio call signal and the phone numbers of two of the skipper's closest relatives. The computer forwards this information to the Coast Guard headquarters nearest the ship's port of call so that duty officers can start mobilizing an air rescue.

The computer then fixes the EPIRB's location. Because it works with sets of digital data, it winds up calculating two Doppler positions—a *true* image, which is where the EPIRB actually is, and a *mirror* image. The computer weighs the composite images by percentage of accuracy, discerns the beacon's true position and forwards the coordinates to the Coast Guard station closest to the emergency. Ninety-five percent of the time, the computer determines which is the true Doppler image in less than five minutes.

This time, it was not certain.

When Bob Doyle activated the 406 EPIRB aboard the *La Conte*, SARSAT-4 happened to be very low on the horizon. Orbiting the poles at 4.37 miles per second—15,372 mph—the satellite received only a short, incomplete burst of data before it slipped behind the earth.

With little data to work with, the best the Mission Control mainframe could do was offer a *position split*—two possible locations for the emergency beacon.

As the computer saw it, there was a 52 percent chance the distress signal had come from latitude 58°15.5' north, longitude 138°07.8' west. But there was also a 48 percent chance it had emanated from latitude 61°28.3' north, longitude 120°34.5' west. These coordinates were eight hundred miles apart.

To confirm the EPIRB's true position, the computer would need more data. It would have to wait. SARSAT-4 took an hour to complete a pass around the earth. And there was one other problem.

The computer had been unable to identify the ship or its crew; the boat's owner had not bothered to register the EPIRB. If he had, a duty officer in Juneau could have called family and friends of the crew, confirmed that this was not just another false alarm and ordered an immediate launch. It could have shaved fifteen minutes to an hour off the time a helicopter needed to effect a rescue.

That, of course, was no longer possible.

Quartermaster Blake Kilbourne yawned and rubbed his eyes. It was storming outside. He checked his watch: 7:02 P.M. He still had to stand half of a twenty-four-hour watch at the District 17 Rescue Coordination Center in Juneau.

It had been a slow day: a couple of medevacs, a boat that needed a tow and a false EPIRB alarm. Some kid on a docked boat had knocked the beacon out of its magnetic bracket by accident and triggered a Mayday. There was no harm done. Kilbourne had radioed the vessel and aborted a rescue.

Lieutenant Steve Rutz, sitting beside him at the console, asked him, "How you holding up?"

"I could use a nap."

"Go ahead."

"You sure?"

"Go ahead. I'll call you if I need you."

Kilbourne took their dinner plates to the kitchen, rinsed them and the utensils, stowed everything in the cupboards and went to the bunkroom. He was unfolding his sleeping bag when he heard the zipping noise of the SARSAT-dedicated printer coming from the control room.

"Hey, Steve?"

"Yeah?"

"You got that?"

"I got it," Rutz called out. "Get some sleep."

Rutz ripped off the bulletin and scanned it:

406 FIRST ALERT/SIT 174/FOR CGD17
MCC TRANSMIT TIME: 31 0353 JAN 98
DATA FROM SAT/ORBIT: S4/48220

What he read next stopped him: the computer was giving two possible positions for the same distress beacon. That's weird, he thought. Maybe a corrective message will come over. He waited.

Nothing.

Rutz took the bulletin to the chart table and plotted both sets of coordinates on the Alaska map. The positions were eight hundred miles apart. One was close to the Aleutian Islands. The other was ninety miles west of Cape Spencer, on the Fairweather Grounds.

It could be a false alarm, he thought. Let's see if I can't call the boat owner or the family of the crew. He read the message again, under the words BEACON DATA:

LONG MESSAGE: N/A

EMERGENCY CONTACT: N/A

BEACON ID IS NOT IN REGISTRATION DATABASE

There was no saying exactly how bad conditions were in the gulf. There wasn't a single data buoy out there and only one in all of Alaska—near Kodiak. Unless a commercial vessel radioed in data, calculating wave heights, wind speeds and barometric pressure would be educated guesswork.

However, he had some general information, and the news was not good.

The National Weather Service was reporting twenty-foot seas across the Gulf of Alaska, thirty-five-knot winds. If there were people in the water, their chances weren't good. Water temperatures in the gulf were between thirty-seven and thirty-nine degrees. In water that cold, a two-hundred-pound man in a survival suit had an 83 percent chance of lasting two and a half hours, Coast Guard studies showed. After that, the chances of survival plummeted, especially if wave heights were more than twenty-five feet. The higher the seas the faster a person burned body fat and the less time it took for hypothermia to set in.

Satellite imagery showed that on January 28, a huge low-pressure system centered south of the Aleutians had spun off several small storms. These storms had merged and formed one deep *bomb*—a tightly packed, cyclonic storm with powerful gusts at its edges. The bomb had tracked south and east and then, on the twenty-ninth, had swung north. Moving toward the gulf, it was fed by ocean winds and swells that had originated as far south as Oregon. Such a long fetch—the distance over which wind could blow to generate swells—was troubling; the swells had plenty of room to grow in, and time and space to energize the seas.

It was a tough call. Rutz had conflicting weather information, two equally possible EPIRB positions and no information on the vessel's port of call, its crew or its owner. Screw it, he said to himself. I'm playing it safe.

At 7:13 P.M., he issued an Urgent Marine Information Broadcast.

A 406 UNREGISTERED BEACON HAS BEEN DETECTED BY SATELLITE IN THE GULF OF ALASKA, APPROX 50 NM WEST OF CROSS SOUND IN POS 58-15N, 138-08W.

MARINERS IN THE AREA ARE ASKED TO KEEP A SHARP
LOOKOUT FOR SIGNS OF DISTRESS, ASSIST IF POSSIBLE,
CHECK THEIR OWN EPIRBS FOR ACCIDENTAL ACTIVA-
TION, AND MAKE ANY REPORTS TO THE NEAREST COAST
GUARD STATION. SIGNED US COAST GUARD JUNEAU
ALASKA.

Three minutes passed, then five, then ten. There was no response. Rutz reached for the phone and dialed Air Station Sitka, the emergency number.

T H I R T Y - T H R E E

Right as the kid hit the jumper and the crowd went bananas, his pager went off.

Betty Jo turned to him.

"Ted?"

"It's all right," Ted LeFeuvre said. "Just one of my pilots." He read the message and the time it was sent: 7:41 P.M. "I'll be right back."

He made his way down the bleachers through the crowd and out to the hallway. Turning on his cell phone, he walked over to an outside door to get reception and dialed. He heard Dave Durham's voice.

"It's Captain LeFeuvre, Dave," he said. "I just got your page. What's going on?"

"Captain, the station got an unregistered 406 EPIRB alert about ten minutes ago."

"Where?"

"We're not sure yet, sir. It seems that mission control got a fifty-fifty split on the first satellite deposit. One of the solutions is eight hundred miles out, near the Aleutians. The other is a hundred and fifty miles northwest of here—the Fairweather Grounds. We figured we'd launch on the grounds, being that it's closer."

"Smart," Ted LeFeuvre said. "Are there people in the water?"

"We don't know."

"Weather?"

"Well, not too bad. Observation tower here is saying thirty-knot winds and, I think, twenty-foot seas. Aviation weather is calling for lots of *snain*."

"Who's flying?"

"Lieutenant Adickes and Lieutenant Molthen, sir. They're taking a flight mechanic and a swimmer—Witherspoon and Sansone, I think."

"How are the helicopters?"

"All three birds are in good shape. Adickes is going to use the 6018."

"Good," Ted LeFeuvre said. "Tell them to go ahead. You're at home?"

"Yes, sir."

"Are you going in?"

"Right now."

There was something about the way Durham had said it that made the back of Ted LeFeuvre's neck prickle. He sat down next to Betty Jo. There was a time-out on the court.

"What's going on?" she asked him.

"We just got an EPIRB signal."

"What's that?"

"A distress call from a ship. We're launching a helicopter right now to investigate."

"Sounds exciting."

"It's probably nothing."

The teams came back out on the court and play resumed. It was a close game. Betty Jo hollered and yelled when her son scored. The crowd was very loud. A big groan went up each time Sitka missed a shot and a thunder of applause when the home team scored. Ted almost didn't hear his pager go off again. He looked at the message and the time, 8:02 P.M.

"I'll be right back," he told Betty Jo.

Out in the hallway he dialed the number for the ops center at the air station. He heard Dave Durham's voice.

"Sorry to bother you again, Captain," Durham said.

"Not at all. What's the latest?"

"Well," Durham said, "Bill Adickes launched ten minutes ago and he's asking for cover."

"Bill Adickes?"

"Yes, sir."

Adickes wasn't the kind of guy who asked for cover. Dan Molthen didn't either. Together, those two guys had close to five thousand hours in the H-60. They didn't rattle. Why would they want an escort plane?

"Well," Ted LeFeuvre said, "let's get them cover."

"Yes, sir."

"Call up Kodiak."

Air Station Kodiak had C-130 turboprops on the ready. It would take roughly forty-five minutes to launch one, Ted LeFeuvre thought, and another two and a half hours for the plane to make the Fairweather Grounds.

"Actually, Captain," Durham said, "we just got off the phone with Kodiak. They said they're having a real bad snowstorm."

"Great."

"There was an avalanche and snow is blocking the road to the base. The ready crew is having a hard time getting in. The runway's buried, too. They're digging out but it's hard to say when they'll get a plane up."

Ted LeFeuvre said, "Let's launch our own, then. We can get there quicker. And anyway, it's probably just going to be a false alarm."

"Yes, sir."

"We'll be back home before they ever get on scene."

"I'm going to put together a crew," Durham said. "We'll just . . ."

The call was breaking up.

"Repeat that," Ted LeFeuvre said. "Dave, say again?"

". . . won't take too . . ."

"Repeat that. Hello? *Hello?*"

Durham's voice returned. "Captain," he said, "I was just saying that I'll put together a crew and get airborne right away."

"Negative," Ted LeFeuvre said. "Assemble a crew but hold on deck and wait to see how things go. With the tailwind, you guys can get out there quick. If they still want cover later, then you'll launch and give them cover."

"Yes, sir."

"Keep me posted."

He hung up and went back in the gym. He climbed back up through the crowd. Betty Jo smiled at him.

"Everything okay?"

"Sure," he said.

"Ted?"

Ted LeFeuvre was looking in the direction of the court but it was as though he wasn't seeing it.

"Ted, if you need to leave—"

"No, no. Not yet."

The game was very close. The lead changed hands over and over, and the crowd was screaming with every shot, every rebound. Ted LeFeuvre was only half aware. His eyes had a stony look. He stood up.

"Where are you going?"

"I'm calling the station again."

The ops phone rang once when Durham picked up. "Okay, Dave," Ted LeFeuvre said, "what's the latest?"

"I was just going to call you, Captain. Kodiak has been unable to launch. Our 6018 is en route. But, sir, on their last call they reported wind speeds greater than seventy knots."

"*Seventy* knots?"

"Yes, sir."

Ted LeFeuvre felt like he'd been hit in the abdomen with a club.

"Oh."

"You know, sir," Durham said, "I've got Russ Zullick here to fly with me, and we've got a flight mech and a rescue swimmer ready to go. Rather than wait, I'd like to launch now."

Seventy knots, Ted LeFeuvre was thinking. That meant the wind was blowing over 90 mph. Anything above 64 mph was considered hurricane force. What were those crewmen flying into?

"Captain?"

"Yes, yes. Go ahead, Dave. Get going."

"Yes, sir. Thank you."

"No, thank *you*."

Ted LeFeuvre hurried back inside the gym and pushed his way through the crowd and up into the bleachers.

"Listen," he said to Betty Jo. "I need to go to the station. I could take you home now or you can get a ride with Angelina." Angelina was Betty Jo's daughter.

"What's wrong?"

"We're launching a second crew."

"Oh," she said. "We better go, then."

Ted LeFeuvre dropped off Betty Jo and one of her friends and then drove straight to the air station. As he swung the Cherokee into the parking lot he noticed a tow tractor pulling an H-60 out of the near runway. The rotor blades were flapping up and down in the wind.

Throwing open the side-entrance door, he hustled up the stairway and down the hall to the operations and radio center. Behind the desk, a man in a flight suit sat hunched over, talking on the phone.

"Yogi?" Ted LeFeuvre said.

Guy Pearce, one of his pilots, hung up the phone and turned around. Sitting across from Pearce was a female storekeeper who normally worked the ops desk at night. She was on the high-frequency radio.

"Yogi," Ted LeFeuvre repeated, "what are you doing here?"

"Mr. Durham called me in, sir," Pearce said. "He asked me to handle the desk tonight."

"Oh," Ted LeFeuvre said. "What's the latest?"

A second helicopter, the 6029, was preparing for takeoff. Dave Durham was in the pilot's seat. Russ Zullick was copilot. Flight mechanic Chris Windnagle and rescue swimmer A. J. Thompson were going along for the ride. A half hour earlier, District 17 headquarters had confirmed that the distress signal was originating from the Fairweather Grounds, latitude 58°13.8' north, longitude 138°19.4' west. Sitka was reporting a two-thousand-foot cloud cover, twenty- to thirty-knot winds and visibility of six miles.

"That doesn't sound too bad," Ted LeFeuvre said.

"No," Pearce said. "But now listen to what the first helicopter crew said."

Outside of Sitka Sound, Adickes had reported *sustained* winds of seventy-five knots. Visibility was a few hundred yards in blowing hail, snow and sleet. Over the ocean the cloud ceiling was 350 feet; the atmospheric freezing level was below 800 feet.

"Is that what he said?"

"That's what he said."

Nobody will be able to fly over the storm, Ted LeFeuvre thought. Helicopters couldn't last long above the freezing level. The Jayhawk had a deicer, but it only worked on the windscreen and rotors. Any more than a few minutes above the freezing level and the Jayhawk would load up with ice and drop like a Popsicle from the sky.

"All right," Ted LeFeuvre said. "When was the last time we heard from Adickes?"

Pearce paused.

"His 2045 radio guard."

"What?"

"Eight forty-five, sir."

"That's an hour ago."

"Yes, sir."

Ted LeFeuvre could feel his voice rising. "Get Juneau on the phone. *Now.*"

"Yes, sir."

"I want to know what's going on out there."

Coast Guard aircraft were required to keep a radio guard every fifteen minutes. That was one of the rescue swimmer's in-flight jobs—to keep radio contact with the Rescue Coordination Center in Juneau. But they hadn't. Why not? When an aircraft missed two consecutive guards, it was procedure for Juneau to alert the launch station. Why hadn't Juneau done that?

Pearce hung up the phone.

"Captain?"

"Yes?"

"The last time Juneau heard from the 6018 was at 2045. The helicopter was weak transmitting. Kodiak was talking to them for a while but lost them."

Ted LeFeuvre looked at the wall clock. It was almost ten minutes to ten.

"When did they take off?"

"Twenty hundred hours," Pearce said.

"And we haven't heard from them since . . . ?"

"Twenty forty-five. That was the last report Juneau had from the 6018. They missed their nine o'clock guard, their nine-fifteen, their nine-thirty and their nine forty-five."

Oh, my, Ted LeFeuvre said to himself. Then, to Pearce: "You launch Kodiak. You get Kodiak on the phone and tell them we want a plane over-head. I don't care *how* long it takes them."

"Yes, sir."

All right. There it was. A bird was missing. It had lost comms an hour ago. It was flying through whiteout at night 150 miles offshore with a low ceiling and freezing level. In seventy-five-knot winds. Without comms. And I was watching kids throwing a basketball in a hoop.

"You get through yet?"

"I'm on hold, sir."

It could be worse. Okay, one of your birds has lost comms near the Fairweather Grounds. Stay cool. Think. That area is notorious for poor HF radio transmission and reception. It has always sucked in radio waves. It's known as a "dead spot." Could they have gone down? Don't you dare think that. That is one thing you better not think about. But this could be some-thing bad. Like what happened last summer in Eureka when that H-60 flew straight into the water. Sank like a crab pot. When they pulled it up from two thousand feet of water, they found all four airmen still in their safety straps.

All right, what are you going to do now? Keep thinking ahead. You were always good at that. So do that. Our second helicopter is taking off right now. Good. Dave Durham is flying it. He's one of the best pilots you've got. And Russ Zullick is a darned good one, too. Think about them. Think about your next move. How many pilots have you got left? Adickes, Molthen, Durham, Zullick are already flying. Taylor, Buchanan, Gebele and Bellatty are out of town. That's eight pilots right there. What about Jack Newby? No, he and his wife had a dinner party and he's been drinking. Let's see. Guy Pearce is here. Who else have we got?

"Captain?"

"Yeah?"

"Kodiak says they still haven't gotten a plane out yet. They got hit by a blizzard and an avalanche of snow—"

"I know all about that. Tell them to launch when they can."

Turning, raising his voice as he spoke into the receiver, Pearce said: *"Launch when you can, then. Right."*

He hung up.

"Kodiak is working on it, Captain. They'll launch as soon as a runway's cleared."

"They said that?"

"Yes, sir."

"Well," Ted LeFeuvre said, "I guess that's the best we can do."

BOOK FIVE

On the launchpad, wind speed was twenty-five knots. No sweat, Bill Adickes said to himself This helicopter is a flying tank. She'll handle twenty-five knots like a summer breeze. He noticed something on the dashboard.

"Oh crap," he said.

Dan Molthen, who was running through his instrument checklist in the pilot's seat, looked over.

"What's wrong?"

Adickes bit his lip. "Looks like the NDB is down."

"No."

The NDB was the aircraft's nondirectional beacon. A tiny needle on the control panel, it homed in on the airport's radio station and made navigating in and out of Sitka Sound, even in poor visibility, a snap. By setting a 2-0-0 bearing off the NDB signal and following a straight line, a pilot could fly between an opening in the mountains that surrounded the sound and head straight out to the Pacific.

"Let me check it again," Adickes said.

Damn the luck, he said to himself. Of all the nights in four years of flying here for this thing *not* to be working. He was flipping through his flight manual now. Nine times out of ten this is a button-pushing error, he thought. This H-60 has got very complicated displays with multifunction buttons, and damned if I can remember half the time what all of them do. Maybe I accidentally disabled the NDB. No, it's still on. Maybe I forgot to turn on the radio. No, that's on, too.

He slapped shut the manual.

"It's broken."

"Shit."

Adickes peered out the windscreen. A slanting rain like a swung curtain of silver beads was raking the runway. He couldn't see the mountains. He couldn't even see the end of the runway.

"Well," Adickes said, "we've got radar. If we need it we can fly by the GPS. I say we go for it."

"I'm game."

Molthen clicked a button on the intercom so he could talk with the crew. "Crew report," he said. "Ready for takeoff?"

Through the headset he heard the deep voice of his flight mechanic, Sean Witherspoon. "Crew's ready."

"Here we go," Molthen said.

He pulled back on the collective, the control stick on the floor to his left that fed fuel to the main rotor head, and the Jayhawk lifted slowly up and hovered ten feet off the runway. Molthen and Adickes scanned the LED readouts, flight controls and tort, or power, gauges.

"Tort's eighty percent, heading instruments and flight controls feel normal," Molthen said. "Radar check?"

Adickes hit a switch and the radar screen lit up. "Radar's looking good," he said. He nodded at Molthen and flipped down his night-vision goggles.

"Okay," Molthen said, *"I'm on the go!"*

He dropped the nose five degrees, pulled power and they were off, sweeping along the runway and then fast-climbing up and away from the dimming airbase lights and into the pitch blackness above Sitka Sound.

"Okay, Dan," Adickes said to Molthen, "let's take her up to three hundred feet."

"Taking her to three hundred," Molthen said.

Just then a gust yanked the nose of the aircraft sideways. Molthen wrenched the cyclic, the floor stick that controlled the Jayhawk's lateral movements, hard left.

"That wasn't very nice," he said.

"I don't like it," Adickes said.

"What?"

"The way these winds are coming over the mountains south and east of the sound. Look how they're hitting the mouth of Silver Bay."

He fed Molthen a radar heading that would give them a wide berth of Mount Edgecumbe, the highest, snow-topped volcano at the mouth of the sound.

"Keep an eye out for Edgecumbe," Adickes said. "And stay on this heading."

"Staying right on it," Molthen said.

Climbing, the rain so thick it seemed like flying through a waterfall, they started feeling a heavy tremor in the joysticks. For two minutes in the air that's a lot of wind, Adickes thought. And a lot of water. How much water can be sucked into the intake of an engine before it flames out?

"Goddammit," he said. "I can't see shit. You see Edgecumbe yet?"

"Not yet."

"Christ," said Adickes.

In the rear cabin, Rich Sansone, the rescue swimmer, was already on the radio. He turned the volume up on his HF headset to hear over the hail and wind. "Comms Center, Juneau," he said. "Comms Center, Juneau. This is Rescue 6018."

"Comms Center here."

"How do you read me?"

"Loud and clear, Rescue 6018."

Sansone said, "We have four people on board, repeat, oh-four P-O-B. Our destination is latitude 58 degrees 13.8 minutes north, longitude 138 degrees 19.4 minutes west. We have an ELT hit at that position. Our mission is search and rescue. We are airborne at this time and we are en route to that position. Request you take our radio guard."

"Roger, 6018. Accept your radio guard at 0500 Zulu. Report flight ops and position every one-five minutes."

"Roger," Sansone said. "Readback correct. Rescue 6018 out."

In clear skies with no wind and a NDB to set his outbound course by, a pilot might have leaned back leisurely and steered the aircraft by guiding the hydraulically powered cyclic and collective sticks with the tips of his fingers. With the helicopter ramming into oncoming gusts, Molthen's hands were clamped to the cyclic and collective, his back arched as he leaned forward, his feet pumping, easing and pumping on the floor pedals, his eyes darting around the console.

There was nothing to see through the windscreen but black, so he focused on his panel instruments: the airspeed indicator, the vertical speed indicator, the altimeter and the attitude indicator, a big, blue gyro ball that told him how level the aircraft's nose was. Molthen had to keep that gyro even. If the gyro floated up even a little, a wind would snag the aircraft by the beak and send it pitchpoling backward.

"See anything yet?" Adickes asked him.

Molthen flipped down his goggles and glanced out his side window. Through the goggles everything looked grainy green and black, like the screen of a TV with bad reception. He spotted a fuzzy line that made a conelike silhouette.

"There it is," he said. "Mount Edgecumbe. Over there. See it?"

"No."

"Well," Molthen said, flipping up his goggles, "it's there."

"Okay, well, let's get the hell around it and outside on the ocean," Adickes told him.

"Roger."

The nose jumped and swung, worse than it had after liftoff, and Molthen was wrestling the sticks now, pushing and pulling them forward and sideways, pulling and pushing, struggling to keep the helicopter level against the shifting, intensifying stream of wind. Again and again he corrected the heading. The aircraft kept twisting in a series of steady jerks, the cyclic and collective bucking in his hands.

The helicopter moved past the cape.

"We're outside," Molthen said.

"Good."

Then the gust hit.

It was like being smacked by an enormous flyswatter. The Jayhawk went over hard and everyone went with it. Molthen was pinned to the side door. He tried to push off his shoulder but it did no good. He slammed his boots into the pedals and struck back hard on the cyclic.

"Good Christ!"

The Jayhawk was spinning, wobbling.

"Compensate! Compensate!"

"I got it!"

Molthen swung the H-60 back into the sledgehammer of air, then spun it back around so the winds would grab the stabilizer. They took off like a rocket.

"Whoa," he said. "We're moving now!"

"Aren't we?"

It felt as though they were hurtling through a wind tunnel. Adickes pulled up the hover page on his main display. Until that moment, the anemometer had indicated twenty-five knots of wind speed. It was now showing thirty knots.

Then thirty-five.

Adickes's eyes skipped over the other instruments, then stopped again at the wind-speed indicator.

It read forty.

And then forty-five . . .

He couldn't take his eyes off the gauge.

. . . fifty, fifty-five, sixty . . .

. . . *Jesus* . . .

. . . sixty-five, seventy, seventy-five . . .

Seventy-five knots of wind? That can't be, Adickes said to himself. That's 110 mph of wind. And these are not gusts. The computer is measuring *average* sustained winds.

He checked their ground speed. Molthen was giving the helicopter enough gas to fly 150 knots an hour. But because of the tailwind, their true airspeed was 225 knots an hour—275 mph.

This is truly something, Adickes thought. Truly something. But what is this tailwind going to be two and a half hours from now? It's going to a headwind. And it's going to make us burn a lot of fuel trying to get back. So start using your head and be smart with your fuel. Calculate your fuel-burn rate. And get some help.

"Dan," he said.

"What?"

"We need to ask for air cover. We need a C-130 out here with us."

Molthen only nodded his head. He had already reached the stage of the fight where he needed to save his breath. He was harnessed to the controls. It looked as though he was strapped to a mechanical bull with control sticks and floor pedals. He did not stop moving.

"I'll call for it," Adickes said. "You fly. Watch the water. I'll do the navigating."

Adickes turned up his HF radio and raised Sitka. The transmission was scratchy. But he was able to report his position and flight conditions and make his request for C-130 air cover understood.

He signed off.

"Okay," Adickes said. "I told them to ask Kodiak for a C-130. I hope they're getting cover."

Molthen nodded.

"Man, we really got a tailwind," Adickes said. "We're going to get there pretty quick."

Molthen only nodded.

The Jayhawk came with a large, flat stabilator tail that normally helped smooth out turbulence. But wind shears were spiking that tail and making the rear of the helicopter slew about and bounce as though they were riding a slow-moving jackhammer.

"Say, Dan," Adickes said, "how about trying to find some clearer air?"

"Okay."

"Let's go up to four hundred feet."

They rose to four hundred, then five hundred, and on up to eight hundred feet. But it did no good.

"Hey," Adickes said, "we're getting ice on the frame. Better take us back down to two hundred and fifty feet. Maybe we can get a visual of the ocean from down there."

"Down to two hundred and fifty," Molthen said.

But they couldn't see much, not even with the night-vision goggles. The goggles could only enhance existing illumination. Outside was pitch-black; there was no light to enhance.

They heard Sansone call out, "Who's flying this plane?"

"What, Rich?"

"I said, what are you guys doing up there? We're trying to get through our checklists. And we can't. Not with you guys bouncing us around so much."

"Hey, you know something, Rich?" Adickes said. "It's going to be like this the whole way out."

"Can't you do something about it?"

"No, Rich, we can't," Adickes said. "It's a bad, bad night, okay?"

"Are we flying out of trim?"

"Yes, Rich. We're flying out of trim." Nobody liked to fly in one direction with the nose of the aircraft angled slightly off. But Adickes knew what

Molthen was doing. He was trying to save gas by allowing the tailwind to push them out to sea while at the same time keeping on course to the EPIRB position.

"I hate flying out of trim," Sansone said.

He and Witherspoon were strapped in side by side along the back wall. Witherspoon kept his head back and his eyes shut. The vertigo had started as soon as they had rounded the cape. He was not sure what was up or down. Everything was moving too fast. Sansone heard his labored breathing.

"Sean, try to relax," Sansone said.

"I'm okay," Witherspoon said. "I'm okay."

"Breathe slowly."

"There can't be anybody out there," Witherspoon said.

"Slowly."

"There can't be anyone. Who'd be out in this shit?"

"Nobody's out here. That ELT is nothing. Probably just a false alarm."

"That's it. Somebody knocked an EPIRB overboard. We'll get on scene and it'll be nothing."

"Breathe slowly."

Sansone checked his Luminox diver's watch. It had been fifteen minutes since his first call to Juneau. He leaned forward and turned up the HF radio.

"Comms Center, Juneau," he said. "This is Rescue 6018 on zero-five megs."

He waited.

"Comms Center, Juneau, this is Rescue 6018 on zero-five megs, over?"

"Yes, Rescue 6018 . . . this is Comms Center, Juneau . . . request flight ops and position . . ."

Sansone read off their latest GPS position and confirmed that flight operations were normal. The operator read back the position.

"Readback correct," Sansone said. "6018 out."

Bill Adickes had been listening to the chatter in the cabin and trying to calculate BINGO, the point at which the H-60 would not have enough fuel to return to base.

The EPIRB is 150 miles offshore, he thought, and we're going to be there in fifteen minutes with this tailwind. How much fuel should I leave us to get back? How strong is the wind going to be? Sixty knots? Eighty knots? If this is a fast-moving hurricane the wind could swing around and present me with a whole different return scenario. Well, I better not cut it close on fuel. Not tonight. These winds are too unpredictable, coming at us from all angles. How is it that they're doing that? Our approach into Sitka is going to be tricky. That crappy NDB. Why did it have to go kablooey tonight? And these damned computer keys. How am I supposed to punch these things in the dark in a shaking aircraft?

Adickes said into his mouth set, "Hey, Rich, how's your boat doing?"

"Sir?"

"Your boat. You know. Weren't you working on it or something?"

"Yeah, I am. I mean—I was. It's fine now. Fine."

"That's good."

There was a silence and then Sansone said, "How's yours?"

"Good."

"Well, that's good. That's good."

Dan Molthen cleared his throat. "Hey, Sean," he said. "I hear you just got married."

In a croaking voice, Sean Witherspoon answered, "Last week."

"Gee, I'm sorry to hear that."

"Thanks."

"How bad was it?"

Witherspoon smiled, opened his eyes, then closed them again. "Not as bad as people said it would be. But still pretty bad. How about yours?"

"Terrible. Still getting over it."

"I'll bet, sir."

"Okay, guys, listen up," Adickes said. "Here are some possible scenarios we might encounter tonight. We might have people in the water and—"

"There's no way in hell anybody's out here," Witherspoon said.

"Amen," Sansone said.

"Bill," Molthen said, "I hope there's nobody out here. Not tonight."

"I hope not," Adickes said. "But you never know. We might have a vessel that's sinking. We might have to pull people off the deck. We might have to do a rescue-swimmer deployment. Or a pump drop. I want you guys to be thinking about all these things as we get closer. And as we get closer, we will discuss each scenario before we go into it. Okay?"

"All right."

"I want you all to talk to me. If I ever go too fast for you, I want you to just say stop. We have time. The five minutes we spend talking about it will save us fifteen or twenty minutes down the road trying to fix the cable we got wrapped around the mast, or God knows what. Let's talk about things before we screw something up."

They were hurtling farther out to sea all the time and the beating they were taking was only getting worse. There was the roar of the hail and sleet on the windscreen and every few moments a gust would cuff the helicopter to one side and then another would slap it back the other way. Everything not pinned down was rattling and jumping about—the Velcro straps and checklist binders and maps and cords—and Molthen was still arched against the strain, his arms and legs a blur, and yet fastened to the aircraft as if he had hooked a giant fish, unable to take even a moment's break from the controls. He was holding the aircraft on course but it was draining his strength and drying his mouth out.

"Want some coffee?" Adickes asked him.

Molthen shook his head. "What do you think we'll find out there?" he asked.

"No idea," Adickes said.

"Wouldn't it be great if we just saw something right away and solved this mystery?"

"Let's just wait and see," said Adickes.

Sansone checked his wristwatch. Another fifteen minutes had gone by. It was time for another radio guard. He turned up the volume on his headset.

"Comms Center, Juneau," he said. "Comms Center, Juneau. This is Rescue 6018."

No response.

"Comms Center, Juneau, this is Rescue 6018. Over."

Static.

Sansone switched to his backup frequency, Channel 16, and tried again.

Nothing.

I'll try Kodiak, he said to himself. It's 750 miles away. But who knows, maybe it'll work. "Comms stay, Kodiak . . . Comms stay, Kodiak . . . This is Rescue 6018. Do you hear me?"

Silence.

"Mr. Adickes," he said. "I think I've lost my radio guard with the comms center in Juneau. Air station Kodiak does not respond either."

"Keep trying, Rich."

But it did no good. Sansone began making calls in the blind.

"Mr. Adickes," he said. "I can't get anybody."

"Did you try the HF?"

"Yes."

"Well," Adickes said, "forget it then. Don't worry about talking to anybody anymore."

As he turned down the radio volume, Sansone felt a swift tightness in his temples and a crawling of skin on the back of his neck. They were very much on their own.

T H I R T Y - F I V E

Down in the raging sea, five men were making their fight. It was hard to find dry air to breathe. The wind peeled their eyelids up and the salt water burned their eyes and seared the insides of their throats and nostrils. It was coming so hard at them that they could not keep the water out of their stomachs long, and every few minutes one or another of them would retch it back up. Their lips were cold to the point of painlessness; they would not feel a fist slam their mouths. With each wave, each gust of ice-laden wind, the water got through their shivering lips, stuffing their mouths and throats and stomachs with more of the tooth-chilling, tongue-burning seawater.

The five men were bobbing in a circle. A nylon rope still held them loosely together, but the lifeline was coming loose around their waists. When they came to within arm's length of one another, each man reached for the other and clung fast, bowing their heads and praying and cursing the sea for all of her wickedness. When they came apart they thrashed madly, frantically, calling out to one another between gasps.

The combers came faster now, each no more than ten seconds apart. The bigger ones came with a whooshing, earsplitting roar. The men screamed but their screams were no match for the noise around them, one wave lifting them bodily up and up and up, feeling the water before them and behind them and inside of them. Then it was as if they were outside of themselves, floating, and then they were propelled down and down and down, the wave rolling them, their legs going over their heads, their lungs aching for true air, their bellies full of the salt water, the most primitive quadrant of their selves trying simply to stay alive, as they tumbled, hopelessly, through the universe.

And then they would feel themselves sliding into the belly of another wave.

From somewhere deep inside his head, Bob Doyle heard somebody weeping, crying out his name.

"*Help me, Bob!*"

Bob Doyle tried to tune in, to find focus.

"*Bob!*"

His throat raw from salt water, Bob Doyle answered weakly, "I'm here. I'm here."

"I can't keep my head up!" It was Mark Morley.

The lifeline was so loose it had gotten tangled around his knees. Bob Doyle kicked and twisted and maneuvered until he had it up around his waist again. He yanked on the rope, drawing the skipper to him.

"I got you," he said.

"I can't feel my legs," Morley said. "I can't feel my legs."

"Don't worry. I got you."

"I can't feel my legs." He was shaking and sobbing. "Oh my God. I'm losing my legs."

Hearing a deep, swishing gargle, Bob Doyle swung his head. Another curler, snarling and foaming at the lip, was rising over them.

"*Deep breath — now!*"

It was hard to know how long they were down. It seemed — forever. Bob Doyle was about to give in and breathe the ocean when he heard the wild shriek of the wind and saw the hood next to him.

He grabbed it.

"Mark!"

Morley's response was faint: "Hold me, Bob."

"I got you."

Then he heard Gig Mork shout, "Where the hell are you guys?"

"I'm here!" Bob Doyle yelled back. "With Mark!"

"Fuck! I *lost* him!"

"Lost who?"

"He was on my chest. We went under. And now—"

"Who?"

"Hanlon, goddammit!"

"Dave?"

"*Shit!*"

"Where's Mike?"

"*MIIIIIKE!*"

A hard, hoarse voice came out of the blackness. "I'm here. I'm here."

"Is that you, Mike?"

"This sucks!" Mike DeCapua yelled out. "Oh, fuck me. Fuck me. There's something wrong with my legs."

"Chrissake," Mork said. "Get the hell over here!" Mork pulled DeCapua up on his chest and leaned back, floating. "You sniveling son of a bitch! Hold on to my arm!"

"Thanks, Giggy."

"Keep your fucking yap shut."

"Where's Dave?" Bob Doyle asked. "Dave!"

They called out his name, but got no response.

"Dave?"

"Dave! *Quit fucking around!*"

But there was no answer. They kept calling out but it was like trying to shout over a passing train during a downpour. Wave after wave drove them under, yet as soon as they surfaced they shouted their partner's name.

"Where is he?"

"Dave!"

"He's got to be here."

"You think he's underneath us?"

One of the buoy balls was missing. It was the buoy DeCapua had tied to Hanlon's waist. Mork stuck his face beneath the water and tried to feel around for him. He even tried to dive underwater to look. But it was no use. His inflated collar held him to the surface.

"Hey!" Bob Doyle shouted. "The ball! There's the buoy ball!"

"Where?"

"Right there!"

Five yards away off, glinting in the flash of the EPIRB, bobbed an orange buoy. It did not fly off with the lash of the wind. It simply bobbed on the surface, heavily. The guy has got to be attached to it, Bob Doyle thought. Otherwise that ball would take off like a shot.

"Dave!"

They watched the buoy come closer, then swing away, then come closer again.

"Dave! Is that you?"

"Giggy, swim to it!"

A wave avalanched over them. Then another, then several more. Each time they came up for air, the orange ball was a little farther away.

"Mark," Bob Doyle said to Morley. "You gotta help me, buddy. We just gotta swim to get Dave. Come on. Move your arms a little. That's it."

The buoy was still in sight. But even as Bob Doyle kicked and thrashed with his one free arm, the buoy kept sliding farther and farther away.

"C'mon!" he barked at Morley. "Don't stop! That's Dave over there!"

He was hollow-sick in his chest and stomach from the effort and yet he fought and fought the current until he thought he would pass out. When the sick feeling lessened he started kicking and thrashing again but it did not bring the buoy ball any closer. It occurred to him suddenly how cold the water was. His legs were numb. Another wave pummeled them. The men popped up, blind and spitting, and Bob Doyle wiped his eyes with his mitten and floated in the icy water and let the current take him along. The buoy ball was out of sight now.

Sleet glazed the windscreen of the Jayhawk. Even when he put his night-vision goggles down, Dan Molthen could not make out the ragged ocean from three hundred feet.

"How much longer you think we are from scene?" he asked Bill Adickes.

"Anytime," Adickes told him.

"We've already passed the fly-to coordinates," Molthen said. "I don't think anyone's out here."

"Well, if there is, they must be drifting fast," Adickes said.

"I'm engaging the deicer." The helicopter was going to burn more gas with the deicer on, but he had no choice. "And let's get that direction finder on. See if we can pick up the EPIRB's signal."

"I'm working on it," Adickes said.

He switched on the direction finder and typed the frequency of the distress beacon into the computer. If the finder detected a 406-megahertz signal it would emit a warbling sound. The needle on the dial would swing and point in the direction of the transmitting beacon. If they overflew the EPIRB, the needle would swing around to the six o'clock position.

"Finder is on," Adickes said.

"Anything?"

The needle didn't budge. It still pointed ahead, to twelve o'clock.

"No," Adickes said. "Not yet."

They had been airborne forty-one minutes and not for more than three seconds had Molthen taken his eyes off of his attitude indicator or the altimeter. I can't let the nose jut up for even an instant, he said to himself. It would be nothing for a gust to catch us by the nose and flip us back into the water.

He was giving it gas only occasionally now to keep the aircraft slightly off trim with the wind. The tailwind was whisking them along, farther out to sea, but Molthen was still working the aircraft exactly as he should and delivering each time Adickes asked him to do something. As he saw it, there was no other thing to do. But he was growing weary from the constant pitching and slewing and the sweat had soaked his dry suit underneath his outer flight suit and the muscles in his hands were starting to cramp.

He was fighting to keep the aircraft level and on course and he was trusting the navigating and the fuel-burn calculations to his copilot. They had lost radio contact with the world. That was unnerving, but he was not yet worried. He had lost contact with the base in bad weather before. Still, it was not a good sign. I hope there is nobody out here, Molthen said to himself. It's not scary flying this fast. What is scary is the wind. This wind is really pushing us. It is pushing us toward something that probably nobody is ever supposed to see. What happens once you get on scene and have to fly a hover pattern in this wind? And what do you do when it comes time to fly back against it? There can't be anyone in the water on a night like tonight. But what if there are people in the water? Could they be alive? If they are, they're probably half dead from hypothermia by now. They say people go insane from hypothermia. They say people kill themselves from the cold. Just pack it in and drown. Stop thinking like that. Stop it. Keep your mind on these instruments and locate the EPIRB and then take it from there. God, this is awful stuff to be flying in. I've never seen anything this bad.

"The way we're moving," Adickes said, "we're probably going to go screaming over the top of this thing. Just be ready to turn on a dime."

"Sure."

"And eyes out for that EPIRB."

"Right."

"Also, Dan, I think we should probably do a Victor-Sierra search pattern once we get on scene," Adickes said. Doing a Victor-Sierra meant flying in tight, diamond-shaped patterns until they had covered an enormous, circular area around the EPIRB's last reported position.

"I'm fine with that."

"Good. I'll program it into the computer."

The cockpit was jackhammering worse now, the shaking so violent it took Adickes several minutes to punch the search command into the computer. Normally, he would have entered it in less than ten seconds.

As he was struggling to make his fingers hit the buttons he was telling them to, a chortling sound came through his headset, faint at first, and then louder. The direction finder's needle twitched and swung completely around.

"We got a hit!" Adickes nearly jumped out of his seat. "That was a hit!"

"Where?"

"We just passed it!"

Molthen threw the cyclic out and swung the helicopter hard left. But the aircraft kept zooming north and west, farther and farther away from the datum point.

"*Hold on!*"

"*Gas! Gas! Gas!*"

Molthen had the cyclic as far as it would go, but it was as if a giant hand were shoving the helicopter farther out to sea. It took him half a minute just to turn the helicopter; in that time, they had flown ten miles downwind of the EPIRB.

Once he had the nose cocked straight into the wind stream, he fed the engines gas. He gave them enough to accelerate to eighty knots of speed. But the helicopter wasn't moving.

"*Full power! Full power!*"

Now the real fight began. Before, Molthen had only been holding the aircraft level and at a certain altitude while the tailwind drove out to sea and whisked the helicopter with it. Now he was up against the full force of the hurricane—a 100-mph headwind and boomer gusts like the paws of a gigantic bear cuffed them back and forth and side to side like a piñata.

"Okay," Adickes coaxed him coolly. "That's it. We're moving now. Good. We're moving nice and slow. Nice and slow. Don't rush it. Just keep her moving forward like that."

But there was nothing steady about it; it felt as though they were riding a roller coaster car the way the Jayhawk jerked and shuddered, lurched and pitched. Piloting had become a matter of brute strength and Molthen was using all he had in his arms and legs on each stick motion, each push of the pedals, to compensate for the fists of wind pummeling the aircraft.

Adickes kept minding the instruments and coaxing Molthen without overdoing it. Christ, he thought. We're giving an H-60 full power and our ground speed is below seventy knots. How are we supposed to fly an organized search pattern against these gusts? The pattern is going to look like a misshapen bowel when we get done and our accuracy is going to drop. We need to get down closer to the ocean.

He said, "Can you step us down to two hundred feet?" Molthen did not take his eyes off the instruments.

"I can try," Molthen said.

"Not too fast."

They began cloverleafing, dropping down a few feet, swinging the nose across the wind, then bouncing up a few feet before dropping and twisting down lower. Their ground speed slowed to forty-five knots, then thirty knots, then twenty knots. Gradually they came down to two hundred feet. Adickes gazed down through his chin bubble, the curved window on the floor of the cockpit between his legs. With his night-vision goggles on he saw white streaks of sleet and snow, some blowing foam and an occasional whitecap against a black pallet of ocean. Nothing more.

"Closer," Adickes said.

"How close?"

"A hundred and fifty feet."

The warbling sound was stronger now in his headset.

"You're doing fine," he said. He noticed Molthen was having to work extra hard to make himself drop their altitude. He was doing it, but fighting his instincts. "Take us down easy," Adickes said in a calm but firm voice.

Now the sea was appearing in glimpses. He flipped on the floodlights, which threw cones of light down from the aircraft's belly. Shafts of light lit up the snow and sleet in blurry lines. He looked out his side window.

The ocean was so dark it was like trying to spot a glint of moonlight at the bottom of a well. Sometimes he could see boils of foam or the crest of a wave shearing off and snapping away in the wind like a long, white whip. But the seas were not running anywhere, only jumping up and down. It was like the water in his bathtub when his three-year-old, Ryan, was pushing it down and releasing it. No thinking about Ryan now, Adickes said to himself. This doesn't make any sense. Why isn't there a dominant pattern to the currents? Where's the swell line?

"Bring us down a little closer," Adickes said.

"Down a little closer," Molthen repeated.

The helicopter was bucking and swaying but moving in the direction of the EPIRB signal. Now, too, it was easier to pick out the tops of the waves, massive black shapes that bulged and arched themselves up seemingly free of the ocean, and Adickes wondered how any ship—even a thousand-foot freighter—could stay upright in such chaos, and he thought how very fortunate he was to be sitting in the heated, encased cockpit of a helicopter peering down at it. It suddenly occurred to him that at any moment, as the aircraft's commander, he could take over the controls and pull them out of there. He could fly them to shore whenever he wanted. That thought calmed him. Keep your head clear and level, he told himself. Keep it focused. You've got an EPIRB to find.

And then he thought he saw something.

It was a twinkle, a weak point of light—like seeing a lightning bug on a muggy night from several hundred yards. That's no wave crest, he thought. He adjusted his night-vision goggles.

Again he saw the blink.

"I think I see something," he said.

"Where?"

"Take us forward and left."

The next time it took him more than thirty seconds to spot the twinkle of light. That's an EPIRB all right, he thought. Maybe some big ship dropped it. I hope so. I hope this is the end of the case so I can go home and crawl into bed beside Carin and hug her and hug her and hug her.

They pounded on, yard by yard. Twice more Adickes spotted the blink. And then he saw blurs—squiggles, to be more precise, like those in a night photograph of cars driving with their headlights on. There's something down there, he thought. It could be a raft. It could be people in the water. It could be debris from something that fell overboard a ship. It could be the boat sinking. But dammit, there's something there.

Then, as the cone lights from the aircraft's belly brushed the blackness below, he saw pairs of the squiggly lines moving together, from side to side—the reflective tape on the shoulders of a survival suit.

His jaw tightened.

"There are people in the water."

———————

"Don't you *ever* quit whining?"

"All I'm saying is—"

"Oh, shut up," Gig Mork said. "Shut up!"

"But—"

"Shut the fuck up!"

The four of them had managed to hold on to one another since the rogue wave took David Hanlon. Bob Doyle was struggling to keep Mark Morley on his chest. The man felt heavier somehow. He was listening to Gig Mork and Mike DeCapua sniping at each other.

"We should have gone," he heard DeCapua say.

"Shut up!"

"We should've pulled the gear—"

"If you hadn't noticed—"

"—and gone in. Why didn't we?"

"—we're in the fucking mess together!"

"Why?"

A curler with a barrel big enough to carry two Winnebagos slammed them. When they came sputtering up, the first thing Bob Doyle heard from DeCapua was:

"I hate this shit!"

"I hate you!" Mork yelled right back. "I'm letting you go."

"Wait, Giggy—"

"I'm letting you go!"

"No, don't!"

"Well, then," Mork said, "you shut—"

"Okay!"

"—the fuck—"

"Okay!"

"—up! Bob, I'm gonna kill this—"

"OKAY!"

They lost count of the waves. And in a way, they were also losing perspective on their universe. They'd been sucked into so many wave troughs, buried by so many breakers, blinded and gagged by so much wind, sleet and spray that they no longer fully appreciated how remorseless nature had

turned or how horrible their predicament had become. Everything had
gone so cold and bleak, the roar of waves now reverberated round them in
so many dimensions, that it was as if all grace and gentility had withdrawn
from the world. Past and future had no meaning. Time hung absolutely
still. Sight, smell, touch, taste—except, perhaps, the taste of salt and
vomit—were being stripped away, one sense at a time.

And yet as monstrous as the swells grew, as many times as the combers
left them breathing seawater, there was nothing more horrible than the
rogue waves. Some of the swells they were able to stay on top of, and ride
to great heights. Not the rogues. They made no sound, gave no warning as
they stealthily approached, and they crushed them with such absolute
force that when Bob Doyle popped back up through the surface he wondered, briefly, if he had died and his spirit left to drown in the seas for the
rest of eternity.

After one of those rogues, Gig Mork screamed out:

"Where are those goddamned Coasties?"

"Coming," Bob Doyle said. "You'll see a C-130 first." He didn't believe
it when he said it, but hoped he'd been convincing.

"What's that?"

"A big plane."

"Then what?"

"Then comes the helicopter."

He couldn't picture a helicopter flying in such conditions. Maybe a C-
130. But not a helicopter. They'll circle us all night, he thought. They'll
stay with us until morning. Maybe. But what good will that do? We aren't
going to make it until the morning.

He wondered what was going on at the air station at that moment. I'll
bet they thought our distress signal was a mistake, he thought. No one is
coming out in this shit. Would you? Would you fly out here in this shit?
Maybe I would, he thought, but not to save anybody like me.

A wail from Mark Morley startled him.

"Tamara!"

He had passed out for a few minutes, and just as suddenly snapped
awake.

"Tamaraaaaaaa!"

"No, Mark! It's me, Bob."

"Bob?"

"Yeah, it's Bob."

"Oh," Morley said. He was slurring his words now. "I have a kid and a wife and I ain't gonna make it."

"You will."

A wave buried them. When they resurfaced Morley asked him, "Where are they, Bob?"

"They're coming."

"Why?"

"Why what, Mark?"

"Tamara, I love you," and again he began sobbing.

He's losing it, Bob Doyle thought. Say something. Say anything. "Listen, buddy, you're gonna see her tomorrow. Tomorrow."

"Tamara?"

Another wave pummeled them into frigid black. When they popped up, Bob Doyle pulled Morley back up on his chest. The skipper's teeth were chattering, his body jerking. "I hope we're rescued soon, Bob. I'm not doing too good."

"Keep trying, buddy. We've all gotta keep trying."

"I need to . . . my . . . my . . . kid . . ."

Bob Doyle put his lips next to Morley's ear. "You're going to see him," he said. "You're gonna be the first one up in that chopper. You just got to hang on till the Coasties come."

He had begun to shiver himself now that water was getting between the seal of his hood and his cheek. It was as though ice was being dragged along his skin, slowly, down his spine, across his buttocks, down his legs. Each drop that entered his suit seemed to mushroom as it slid down his body and gathered in his suit legs. The sensation of pinpricks around his ankles was fading now; his feet were deadening.

"You . . . you got . . . great kids, Bob," Morley said. This guy's amazing, Bob Doyle thought. He's trying to talk to stay alive. Don't tune him out. Talk to him.

Bob Doyle said, "Our kids will play together someday. You'll see. We'll laugh about this."

"Where are the Coasties?"

"Coming."

"How . . . how long we been . . ."

I don't know and I don't want to know, Bob Doyle thought. He did not answer.

"Bob?"

"What, Mark?"

"I can't . . . see . . ."

Bob Doyle felt a sting of panic in his throat. Was the loss of eyesight on the checklist of symptoms for advanced hypothermia? He wasn't sure, but it might have been.

"None of us can see. None of us." That much was true, he thought.

"I'm dying, Bob."

"Don't say that."

"I sank it," Morley said. "I was the skipper . . . and I . . . sank it."

"Stop talking like that."

"They're gonna . . . sue me," Morley said. "Dave is . . . gone and . . . I'm—I'm right behind him."

"Stop it."

"I'm dying. And I hate it."

"You got to fight. You hear me? Think of your kid. Remember? You're going to have a kid."

Morley's head went forward and his shoulders began shaking. He was sobbing again.

"I hate it," he said.

Just then they saw the light in the sky.

THIRTY-SEVEN

Everything was jumping around the cockpit: binoculars, flight manuals, maps, the pilots. Dan Molthen did his best to read his instruments. But with the helicopter bouncing as it was, the dials were a constant blur.

"How many are down there?" he asked.

"I don't know," Bill Adickes said. "Maybe four, maybe five people in the water."

"I can't believe it."

"There could be more," Adickes said. He flipped up his night-vision goggles. "There could be as many as eight. Give me the binocs." He held the binoculars to his eyes, and then lowered them. "I can't see shit with these things. There's no way I can be sure. But my guess is there's four or five."

The moment he heard Adickes say he had spotted reflective tape, a wild excitement had rushed through Molthen. He still felt it, actually. He could not control the emotion.

"Rich!" Molthen barked.

"Sir?" said Rich Sansone.

"Get dressed out. Prepare to deploy."

"Yes, sir."

Sansone undid his gunner's belt, knelt on the cabin floor and unzipped his nylon bag. He started pulling out his rescue swimmer's wet suit.

Adickes looked at Molthen sharply.

"No, no, no, no," Adickes said. "We are definitely *not* going to do that. We are *not* putting anyone out in this." He turned on his intercom. "Rich," he said, "there is no way in hell you're going out that door. Sit down."

Sansone let go of the bag. "Yes, sir." He took his seat again. Molthen said nothing. He was looping them back around to the last recorded position of the distress beacon. Why did I have to say that? he asked himself. Of course we're not going to stick Rich in the water. What's wrong with you?

"Bill," Molthen said, "I don't know why I—"

"Forget it," Adickes said. He had seen this thing happen to many men under pressure and he had learned how to amputate the emotion before it took over an aircraft. "You just keep flying this thing the way you're flying it."

Molthen said nothing.

"Dan," Adickes said, "we need to start the hoist checklist."

"Right," Molthen said. He took a breath. "Cabin crew," he said over the intercom, "rescue checklist, part two."

Adickes said to Sansone, "Rich, it's about time to let somebody out there know there are people in the water and that we've seen them."

"Mr. Adickes," Sansone said, "we've lost comms."

"I know that," Bill Adickes told him. "Try every frequency. Have you tried eleven megs?"

"Yes."

"Have you tried eight megs?"

"Yes."

"Well, then, go to five megs. Keep calling. Make calls in the blind if you have to. Just because we can't hear anybody doesn't mean someone's not hearing us."

"Yes, sir."

"Dan," Adickes said to Molthen. "I'll keep an eye on the radar altimeter. You just try to get us in a hover over the survivors."

"How high you think?"

"How about a fifty-foot hover?"

"No," Molthen said. "We shouldn't drop any closer than a hundred feet."

Adickes smiled to himself. The man's got his confidence and judgment back, he thought. "Fine. Our drop limit will be a hundred feet. I'll follow your inputs on the controls."

"We're going to get those guys," Molthen said. "I know it."

"Sure," Adickes said. "Only let's not kill ourselves trying."

Sean Witherspoon was at the cabin door with his hand on the door handle. They had just completed the second part of their rescue checklist.

"Ready for the hoist," he said.

"Begin hoisting," Molthen said.

Witherspoon threw open the door to a blast of snow and ice. It stung his face. A chill shot through his body.

"Arrggghh!"

He lowered the visor on the top of his helmet and, grunting, thrust his head out the door. Ice, snow and sleet pelted him. Then he saw it.

Holy shit.

All of Witherspoon's calculations, all of the mental buildup, all of the plans he had made on the flight out were gone. He had not seen the ocean until now. He had been plotting his moves, thinking out the hoist from beginning to end, in careful detail. But now all the moves were gone. They had sailed clean out of his head the moment he made direct contact with the churning, heaving atmosphere.

He hesitated.

Off to his right, in a trough between two enormous swells, he saw two arms, lined in reflective tape, waving. He shut his eyes.

So this is what it comes down to, Witherspoon thought. All of the training and everything you've ever done in your life—this is what it comes down to. One pivotal moment. What are you going to do?

He opened his eyes. His visor had fogged. He flipped it up. As soon as he did, sleet clawed at his eyes and forehead. He wiped his eyes, pulled the visor down. Panting, he said, "I see two guys down there."

"There's at least four, maybe five," Adickes said over the intercom.

"Oh."

"I want a DMB out there right away, Sean," Adickes said. "Let's get that out, now."

A data marker buoy was a radio float that transmitted a 121.5-megahertz signal. They would lock the drop position into the computer and measure the speed and direction of the current by following the DMB's drift. Witherspoon pulled out the float, activated it and hurled it out the jump door.

"DMB's out!"

Adickes recorded the position. "Good. Got it, thanks."

Witherspoon reached a Mark-25 flare down off the wall rack. He set it on the deck near the door. The Mark-25 was a salt-water-activated flare. It floated and shot white light. It would burn for about twenty minutes. Witherspoon peeled the lid off the flare and popped the top. Sleet and snow were blowing around the cabin. It felt like working in a locust storm.

"Okay, sir," Witherspoon said, huffing. His breathing was coming faster. "I got a flare ready."

Adickes's voice crackled over the intercom. It was a cool voice, free of fear. "Whenever you're ready, Sean."

"Flare's away!"

Witherspoon did not see the flare hit the ocean. He had shoved the steel cylinder pitchpoling out the door when the gust hit them. It was a savage gust, well over 110 knots, and it sledgehammered the aircraft. The next thing Witherspoon knew he was halfway out of the cabin, gripping the jump-door frame.

The aircraft was hurtling back and down, nose up and tail pointed toward the water.

"Nose it over!" he heard Adickes shrieking at Molthen. *"Nose it over, dammit!"*

"I'm trying! I'm trying!"

"Nose it over!"

"I'm doing it!"

"We're backing down!"

"I know!"

Then he heard a sound very few airmen in an H-60 have ever heard— the sound of one force being overcome by a greater one, the sound of two General Electric T-700 1,900-shaft horsepower engines spooling *down*. It was an odd, agonizing drone, like the wail of a wounded coyote. Then he heard the stutter of hail on the fiberglass airframe and the moan of wind in the rotor head. The cockpit lights dimmed. In the flickering light everything seemed to move in slow motion: Molthen and Adickes pulling power on the sticks, Sansone slamming into the rear wall, the aircraft tilting until it seemed to stand on its tail. Witherspoon swung his head. A swell was cresting less than twenty feet below.

If a wave catches our tail, he thought, we'll spin into the sea and sink.

"UP!" he screamed. "UP!"

At first, nothing happened.

Then, slowly, grudgingly, the engines responded. The turbines groaned back to life, the cabin floor leveled off sharply, the altitude indicator on the cabin display stopped falling, and he sensed they were climbing skyward . . .

When a second gust caught them.

This one made them twist and sway, gyrolike—yet somehow, they kept rocketing up and up and up, all the time swaying and gyrating in the wind. They went up as high as three hundred feet before the aircraft stabilized.

Adickes pulled up their position. In fifteen seconds, the twenty-one-thousand-pound Jayhawk had been blown backward just under a mile. He and Molthen looked at each other.

"You ready for more?" Adickes asked.

Molthen took a deep breath and nodded his head.

Fifteen minutes of bucking, pitching and scampering about the sky, and the Jayhawk was back over the dimly blinking beacon.

"Let's get another flare out there," Adickes said. "That first one's almost out."

"Roger that," Witherspoon said. They had two more Mark-25s and a Mark-58, which would burn for a good forty-five minutes. He grabbed a Mark-25, popped the tab and heaved it out. The flare ignited downwind.

"That helps," Molthen said. Without the flares, the sea to him was nothing but a great, sweeping blob of blackness, streaked with foam. Now at least he had some reference for the ocean's movement.

"Okay, Sean, let's hoist."

"Roger."

On his knees, wheezing now, Witherspoon clipped the cable to the ring on top of the basket, lifted the cage and, with both hands, pitched it into the wind.

"Basket's away!"

But instead of dropping down, the forty-pound basket flew straight back toward the tail rotor. Witherspoon stared at it, dumbfounded.

"*It's going straight back!*" There was a panicky timbre to his voice that made Adickes nervous.

"Hey, Sean, take it easy."

"It's going straight back!"

"Calm down and pull it in."

Witherspoon threw the winch in reverse and line came steadily in sweeps onto the reel. When he heard the basket rapping the frame alongside the jump door he reached out, grabbed it and hauled it inside.

Breathe, he reminded himself. Breathe and think about what you need to do here. You need to get us forward of the survivors so that when the wind blows the basket it will blow it back to them. Do that.

He started shouting instructions to Molthen. "Forward two hundred!"

"*What?*"

"Forward two hundred—and right one-fifty."

Then:

"Forward and right, two hundred and fifty."
Then:
"Left one-fifty, back one twenty-five."

Under normal conditions, flight mechanics guide pilots to a hoisting position by telling them to fly forward, backward, left or right by distances of thirty feet or less. The closer they get to the survivors, the less extreme the conning commands become. The two work like dance partners, each gradually understanding the other's timing and needs, until they can anticipate the other's next move.

What Witherspoon was asking Molthen to do was so extreme it was almost ridiculous. There was no predominant pattern to his conning instructions, no way for Molthen to discern a trend from them. And the instructions were coming too slow. One second, the survivors would be bobbing in a trough off to the right, at two o'clock, and the next they would be underneath the aircraft, then forward and left of the Jayhawk. They were disappearing from view, cloaked by the seas, and reappearing hundreds of yards from the last position Witherspoon had asked for.

After several minutes, the helicopter swung above the survivors.

"Basket's away!" Witherspoon yelled.

He paid out cable as fast as the winch would allow, hoping for a letup in the wind so that the basket would not trail so close to the tail rotor. On his knees, grunting, shouldering the cable coming off the reel so it would not rub and fray on the door frame, Witherspoon paid out line with a nervous eye. If that basket sails into that rotor, he thought, we're in the water.

When it dropped a few feet below the glinting disk the tail rotor made, he sighed, relieved.

"Okay, basket's going down!"

Actually, it was penduluming beneath the pitching aircraft. It swung back and forth until Witherspoon couldn't stand the cutting pressure on his shoulder and pulled out from underneath the outgoing cable. As he did this, the cable slapped down on the deck and began sawing on the lip of the cabin door and, outside, on the Night Sun, the aircraft's searchlight.

Witherspoon halted the winch.

"Basket's in the water!"

He saw the cage floating for a few seconds and then a wave collapsed over it. The line went taut; the hoist screeched. The cable leaped from his hands.

"*Shit!*"

Through his sleet-glazed visor he saw two pairs of survivors clasped to each other, waving at him, and twenty yards off now, the trailing basket. Witherspoon was six feet tall and could bench-press three hundred pounds. But the basket was making him sweat. He pulled and pulled and pulled but he could not raise the basket an inch. He felt as though he were hooked to an anchor.

"They're not going for the basket," he said. "They're just looking at it."

"Keep trying," Adickes told him.

He raised the basket all the way back up, checked the cable for burrs, found none, waited for the helicopter to circle around and then tossed the cage out again.

Then he knelt, turned his head toward the door and retched on the deck. Sansone saw it.

"Sean, you okay?"

"No," Witherspoon said. He spit, wiped his mouth and turned back to the cable.

He lowered the basket ten more times. He tried dragging it toward the survivors, swinging it to them, dropping it on them. Nothing worked. Every time it looked as though he was making progress, a wave would fold over the basket, the line would go taut and the cable would fly out of his hands.

"Looks like we're getting it closer," Sansone said. He was looking over Witherspoon's shoulder.

Witherspoon slumped back on the deck. "We're not going to do it! I can't! I can't get them!"

Over the intercom, Adickes could hear the flight mechanic gasping for air. Hyperventilating.

"Let's take a break," Adickes said. He handed Sansone a water bottle. "Give Sean some of this and see if you can cool him down. Dan, take us up."

They held an altitude of two hundred feet while Sansone squirted water into the flight mechanic's mouth. Witherspoon threw up on the cabin floor a second time. Adickes kept talking to him.

"Listen, Sean," he said. "I grant you it's horrible weather. Horrible. The worst weather we've ever been in. But if we don't get these people now, with these seas, it'll be a miracle if they survive."

Witherspoon said nothing.

After a few minutes they dropped down to somewhere close to a hundred-foot hover, although it was hard to tell, really, how high they were flying. The radar altimeter was making no sense; at first it told Adickes their altitude was zero, then 100 feet, then 20 feet, then 110 feet.

They dropped their last three flares in an arc around the survivors. It helped, but after ten minutes Witherspoon turned to Sansone and said:

"I can't do this anymore."

"Sure you can, Sean."

Witherspoon shook his head.

"No."

He sat down on the floor, sucking wind, rubbing and rubbing his eyes. They were all water. He could not get them to stop tearing. His lips, knees, fingers, elbows—every joint in his body felt stiff with cold. Sweat ran all over his body.

"I can't see anything out there, man. I can't . . . we can't . . . we can't . . ."

He leaned back against the wall, laboring for air.

"Mr. Adickes," Sansone said. "Take us up. Sean needs another break."

"All right."

They ascended to 350 feet. Adickes made sure to keep the flares in sight. Sansone unzipped Witherspoon's flight vest. He handed him the water bottle.

"How are you doing, Sean?"

"I'm cold inside and sweating on the outside." He gave his partner a pleading look. "I can't get my fingers to move, Rich. My feet don't respond."

"Hang in there."

"I can't."

"Sure you can."

In the cockpit, the pilots were talking.

"This isn't working," Adickes said. "We've got to try something different."

"What you thinking?"

"Let me take the controls. I'll fly the aircraft from the left seat." He paused. "And I'm going to hover using the goggles."

Coast Guard pilots used night-vision goggles while searching for distress beacons from high altitudes, but they hadn't been trained to hoist or
hover close to the water with them on. Doing so was a violation of regulations. Adickes knew it. But in the Marine Corps he had flown many night
missions using the goggles and was comfortable with them.

"I don't know, Bill," Molthen said.

"Look," Adickes said, "I know what 37:10 says in the book. I know we're
not supposed to use goggles while hoisting. But I'm telling you that it's
worked before in getting people out of the water. It's risky. And we're going
to hover real close to the water. But I can do it."

"I don't know."

"Dan," Adickes said, "those flares are going to burn out. And look at
Sean—he's shutting down."

Molthen hesitated, then nodded.

Adickes turned on the intercom. "Gentlemen," he said in his most con-
fident voice. "Listen up." He was going to fly the aircraft from the copilot's
seat. He was going to fly closer to the water, an eighty-foot hover. And he
was going to do this while using the night-vision goggles. Stone silence followed.

"Any questions?"

Sansone looked outside the jump door at the swarming snow and sleet.
Witherspoon had his eyes clamped shut. He was curled up against the wall
in a cold sweat.

"Any questions?"

Nobody in the helicopter said anything.

At first, things improved. The ride smoothed out. The basket went down
faster. And with each hoist attempt, Sean Witherspoon dropped the basket
closer and closer to the blinking strobe.

If we can just keep doing this, Bill Adickes thought, we'll get somebody.
Once you get the first guy up, the battle is won. We've just got to break the
shell.

On their fifth drop, the basket landed no more than fifteen feet from
the strobe. It floated on the surface for more than a minute. But none of
the survivors went for it.

"Why aren't they swimming to the basket?" said Rich Sansone. He was

looking over Witherspoon's shoulder. He shouted out the cabin door:
"Swim! Swim!"

Just then a gust rammed the aircraft and sent it hurtling backward.

"Twenty-five feet from the water!" Sansone shouted. *"Altitude!"*

Adickes pulled full power on the collective and the Jayhawk snapped skyward. They shot up to 125 feet before Sansone said, "That was too close."

Adickes snapped, "I know, Rich, I *know*. I was twenty-five feet from the
water. Okay. That's where we've *got* to be if we want to get those guys in
the helicopter. So chill out."

Sansone went silent.

Oh hell, Adickes thought. "I'm sorry, Rich. Keep talking to me. It's all
right. Keep talking to me."

Witherspoon heaved the basket out again. Adickes could hear his raspy,
labored breathing over the intercom. There were longer and longer pauses
between his conning commands.

"Sean?"

No response.

"Sean!" Adickes shouted. "Talk to me! I can't see the survivors from up
here!"

". . . Forward fifty . . ."

"Go on—"

". . . and right seventy-five—"

"Mr. Adickes?" Sansone interrupted. He was leaning out the jump
door, trying to judge the distance to the water. "Sir, I think I could pull this
rescue off. If you put me in the water I could get those guys in the basket."

"Sit down, Rich," Adickes said. "There's no way I'm putting you out in
this—"

"But—"

"End of conversation, Rich."

Adickes was talking to himself now in the cockpit. Don't panic. It
would be easy to panic, easy to overcontrol the aircraft. Loosen your grip
on the collective. That's it. Stay focused. Keep your parameters small. Stay
close to the water.

He looked at his attitude indicator. Nose looks good. He read the radar
altimeter. Seventy feet. Not too bad. Ground speed? Close to zero knots.
Flares. Where are the goddamned flares?

He peered out the windscreen.

The basket was swinging and twirling, back to the tail, under the aircraft's belly, forward of the nose. The two smaller flares had already flamed out but he could see that the Mark-58 was still burning. It was rising on a swell.

Adickes watched it rise and rise and rise, until it was at the top of his windscreen.

Then it disappeared.

"Hey," Adickes said over the intercom, "the flare just went out."

Where did it go? It can't have blown out, he thought. Those Mark-58s burn for fifty minutes and that one's only been out there for half an hour. Why don't I see it? What the hell's going on?

In the water, the survivors saw exactly what was going on.

They were bobbing right beside the flare, riding the crest of a rogue that was looming over the helicopter. They could only gaze down in horror at the rotor blades of the Jayhawk spinning below them.

T H I R T Y - E I G H T

The second Jayhawk was on the tarmac and the gas was loaded.

"Russ," David Durham said, hustling across the runway, "which seat you want?"

"You take the right seat."

"You don't mind?"

"No, sir," Russ Zullick said. "I'll navigate."

Zullick swung up into the copilot's seat, strapped himself in and started his preflight checklist. He had more flying time in Southeast Alaska than Durham. But the right seat was the aircraft commander's seat, and Zullick was not one to tell a senior officer his business in an aircraft.

Besides, he thought, what does it matter who's got the stick? We're going to fly cover for Bill Adickes's crew. We aren't doing any hoisting.

"Where's Chris and A. J.?" Durham asked.

"There they are."

Looking across the runway they saw two men coming out of the hangar. Each had a duffel bag and a helmet in his hand and both were jogging. On the left was the flight mechanic, Chris Windnagle. The other man, the taller one, was A. J. Thompson. He was a rescue swimmer.

"I'm going to start the engines," Zullick said.

"Go ahead," Durham said.

Zullick reached up, pressed the starter button, waited and then introduced fuel to the igniters. Both engines caught and ran smoothly. He watched their temperatures come up.

"Engaging the head," Zullick said.

The rotors, which had been flapping up and down in the wind, thudded to life, losing the sag of their great weight. Windnagle helped them do a standard review of the searchlights, the Night Sun, the circuit breakers, the flight instruments. Thompson loaded his gear through the jump door and took his seat along the rear wall of the cabin. He turned on his cockpit display unit and pulled up a GPS map. Then he turned up the volume on the high-frequency radio and strapped in.

"This better be good," he said. "I'm missing out on a cozy pillow and a TNT classic."

"What's the movie?" Zullick asked him.

"*Gone with the Wind.*"

"Very funny."

The brake was set. The ice chalks were in. The stabilator was coming down. They heard a hitch pin clank, the growling motor of the tow tractor and then the *clink-clink-clink-clink* of the tractor riding back to the hangar.

Zullick peered out the windscreen. The break wall along the edge of the far runway was bordered in clouds of breaking surf. Another shitty night in Alaska, he thought.

Out loud, he said, "How come nobody ever gets in trouble on a nice, sunny day?"

Durham pulled the night-vision goggles down over his eyes.

"There are no nice days to get in trouble."

They took off to the south with a right turnout into Sitka Sound. It was 9:34 P.M., and nobody aboard Rescue 6029 minded that there was extreme turbulence and pitch darkness in the sound. We're just going to provide cover for another helicopter that's probably on its way back to base, Zullick thought. With any luck, we'll be back on the ground within the hour.

At that moment, a gray Jeep Cherokee sped across the airbase parking lot and swung to a stop outside the operations center. A slender man with thinning, gray hair and glasses jumped out and hurried to the side entrance.

The man was Ted LeFeuvre, the air station's commanding officer. He was sweating.

Up in the operations center twenty minutes later, Guy Pearce was on the phone with District 17 headquarters in Juneau.

"Any luck getting ahold of Rescue 6018?" He paused. "Roger. Let me know your progress." He hung up and turned around. "Captain, Juneau's been calling, Kodiak's been calling and I've been calling. Nothing but dead air."

Ted LeFeuvre consulted his watch: ten o'clock on the dot.

Well, that meant Bill Adickes and his crew had just missed their fifth radio guard. They had lifted off at eight o'clock, and hadn't been heard from in an hour and nineteen minutes.

"We've got to get ahold of them," Ted LeFeuvre said.

"Yes, sir," Pearce said. "You want me to keep trying, Captain?"

"Do that."

What *haven't* I thought of? Ted LeFeuvre said to himself. I just can't believe there are no other options. Come on. You're supposed to be so smart. Think. How do you contact a helicopter in bad weather without a C-130 in the area?

"Sir, I've got an idea."

"What is it?"

Pearce swiveled in his chair. "How about if I try the air-traffic control center at the airport up in Anchorage."

"Anchorage?"

"Yes, sir. I'll ask air-traffic control there if they have any highfliers transiting the area with HF radios. If they do, we can ask the jetliners to tune to our frequency and try to call the helicopter."

"Yogi, that's *brilliant*."

"Let me call Anchorage."

"Go!"

Pearce got right through to a controller at the Anchorage airport and outlined their scenario. He paused and sat back in his chair.

Ted LeFeuvre pulled at his jaw. Kodiak is going to have to get a C-130 off the ground soon, he was thinking. I do not want to lose comms with another helicopter. Our second bird has been airborne ten minutes now. They'll penetrate the outer edge of the storm in less than fifteen minutes.

"Yogi?"

Pearce hung up the phone and wheeled around. His eyes bulged.

"Captain," he said. "Air-traffic control says Alaska Airlines Flight 196 took off a short while ago from Anchorage. It's heading to Seattle. It's cruising at thirty-three thousand feet. And sir, it's passing close to the Fairweather Grounds right now."

Russ Zullick glanced out his door. Ice had been gathering on the side-view mirror since they had cleared the sound. That was fifteen minutes earlier. The mirror was glazed up now, useless.

"We better watch our icing," he said to Dave Durham. He tried to sound casual about it. "We need to keep an eye on it."

"Say again?"

"Icing. We're getting icing on the airframe."

"You watch the icing," Durham said.

He was keeping a cruise altitude of three hundred feet in high turbulence beneath a low, thick cloud cover. Backing winds were shoving the helicopter into a two-hundred-knot sprint. Hail the size of eggs was battering his windscreen and bands of sleet, snow and rain were swinging across his path like slamming doors.

Visibility was so poor that even if he used night-vision goggles Durham would have a hard time making out a C-130 crossing in front of him.

A. J. Thompson, the rescue swimmer, spoke up. "Mr. Zullick?"

"Yeah, A. J.?"

"We're losing comms," Thompson said. "I've already lost Juneau. And now I can't understand what Kodiak is saying."

Zullick adjusted the high-frequency radio on his headset. He heard a garbled, scratchy voice. Then nothing but static.

"Go to five megahertz."

They switched frequencies and set the volume as high as it would go.

"Do you hear that?"

Thompson frowned. "Barely."

The transmission was fading in and out and garbling, yet intermittently, Zullick could pick out a male voice. It was different from the one broadcasting from Kodiak.

Then:

"*Rescue 6018 . . . repeat . . . This is Les Dawson on Alaska Airlines 196 . . . Alaska Airlines 196 . . .*"

More garble.

"*. . . Flight 196 . . . say . . . you are . . . report . . .*"

Well, Zullick said to himself, it's not a C-130. But there's a jetliner somewhere up there. At least we're not alone. Let's see if I can reach him.

"Alaska Airlines," Zullick said. "This is Coast Guard Rescue 6029. Are you reading me?"

The pilot did not answer.

"*. . . Rescue 6018 . . . Say again, please . . .*"

As far as Zullick could tell, the commercial jetliner was talking to the first helicopter. Zullick could not hear the other Jayhawk, only the Alaska Airlines pilot.

"*. . . Say again 6018 . . . Understand you've crashed in seventy-foot seas? . . . Please repeat . . .*"

Zullick's stomach tightened.

"*. . . understand you're in the water? . . .*"

It was all black outside and sleet scratched at the windscreen. So, Zullick was thinking to himself, we're going to be looking for one of our own guys. These Jayhawks don't float. They don't float. Gosh, I hope they got their raft out. Wait. Strobe lights. They've got strobe lights in their vests. They've got strobe lights on their helmets, too. Oh, I hope they got out of the aircraft okay.

Suddenly he pictured a military funeral. The honor guard handing flags to widows. A twenty-one-gun salute. A helicopter flyby. They're your friends, he thought. You live on their street. Dan Molthen. He's got Theresa and Erin and Ben and Shea. Bill Adickes has got Carin and a boy, Ryan. I think she's pregnant or just had a kid. And Rick Sansone—oh gosh, his wife is expecting, too. And Sean. Sean Witherspoon. He just got married last week.

All dead.

"Russ," Durham asked him. "What's going on?"

"Bill and his crew," Zullick said softly, "might be in the water."

T H I R T Y - N I N E

Rich Sansone threw off his radio headset and lunged to the jump door.

Everything outside—the flares, the rescue basket, the survivors, the swells, the sky—had coalesced into one, black mass.

Sean Witherspoon, already at the door, was screaming: *"ALTITUDE! ALTITUDE!"*

Sansone looked up.

The sea was standing over them.

Actually, it looked more like a wall—a wall with a black, completely vertical face, down which cascaded delicate, white ripples. This wave had no curling crest, just a thin, silvery sheen, like moonlight on a sword. It made not a whisper as it moved swiftly and stealthily toward them.

Oh my holy God.

It was a rogue—a fully developed rogue that was going to engulf them in ten seconds or less if they didn't move.

"UP! UP! UP! UP!"

"DO SOMETHING!"

Bill Adickes was already pulling power on the collective; he'd seen the rogue just as Witherspoon had seen it. He'd flipped a contingency switch on the stick to give them extra thrust, and was thinking: If all you do is pull power and you don't pay attention to the helicopter's attitude you won't fly out of this. Get the attitude right. You've got to get that good climb attitude. Not too much, where you go nose down in the water, not too little, or you'll start to back down. It's got to be just right. Once you break the crest you'll be out of the downdraft. Okay. Here comes the wave. Here it comes.

He had the collective as far back as it would go and still the helicopter felt as though it was moving sideways, in slow motion; not up or down, but sideways, as if it were a spider scampering up the face of a toppling wall.

And then the rogue collapsed.

There was a rush of air and in a white-black crashing the sea collapsed just below the belly of the helicopter and spray and foam were entering the cabin with the force of a power hose. Kneeling, blinded by sleet and snow, hearing the turbines groan and feeling the cabin floor roll under him, Sansone thought: *I'm alive.* Then the deck lurched under him and the rear of the cabin rose high up and then dove sharply and he was on his back, the sleet covering him.

The helicopter wobbled upward. Sansone rolled over on his hands and knees. They were free of the wind shear. He felt his helmet. The ICS cord was still in.

He shrieked: *"What in hell are you guys doing up there?"*

"Flying!"

"Goddammit! That wave almost got us!"

"We know! We know!"

"No! You *don't* know! That wave missed us by *five* fucking feet! What the hell are you doing?"

"Saving our asses!"

"Do a better job!"

"That's enough, Rich!"

Sansone leaned out of the aircraft and saw the line, swaying and penduluming in the wind beneath them, and at the end of it, the basket, spinning like an uncontrolled yo-yo. The cable had rubbed on the Night Sun. The searchlight was gouged and dented across its grate. He crawled to the winch. The motor was still good. But the cable had come apart while spooling. The reel was jammed. Birdcaged. Crap, he thought. We're going to have to haul up all that line and the basket by hand.

"Sean!"

In the forward part of the cabin, clinging to the avionics rack, Witherspoon had his head down.

"Sean!"

Witherspoon didn't move his head.

"Sean, help me get this basket in!"

Witherspoon didn't move. His breath was coming in lunges. He was shuddering fitfully. Dan Molthen heard raspy breaths over the intercom.

He asked, "You guys get the basket in yet?"

"We're doing it now," Sansone said. He was pulling the cable up, hand over hand, as fast as his arms would go. I've got to get this inside. Damn basket's moving too much. It could fly right into our tail rotor. Or get pitched up into the main rotor. That would do it. That would do it, all right. There must be more than 150 feet of cable out. God, it's heavy.

A rescue basket weighs forty pounds, but by the time Sansone pulled in the last of the line, it felt five times heavier. He grabbed the basket and tossed it in the corner. It made a dull thunk.

"Clear! We're clear!"

"How's the hoist?" Molthen asked him.

Sansone was catching his breath. "It's, uh, birdcaged." He crawled over to Witherspoon and undid his hands from the latticed rack. Then he pulled the flight mechanic back to his seat. "Mr. Molthen," he said. "Sean isn't doing too good."

"What do you mean?"

"He's—he's out of it," Sansone said. "He's not moving anymore. We're done."

"What's wrong with him?"

"He's done. That's all."

Molthen checked the attitude indicator. They were relatively level. He looked at the radar altimeter. The reading was fluctuating between 330 feet and 390 feet. Snow was building up on his windscreen. He looked out through his side window and caught the blink of the strobe. It was hundreds of yards off to the right and upwind of the helicopter now.

He turned to Adickes.

"Bill?"

Adickes was concentrating solely on the instruments before him, correcting constantly for the lurch-lurch-lurch of the aircraft's nose. When the rogue wave came at them he had shut out all extraneous noise—all noise, really. He hadn't heard the radar altimeter's alarm go off or Molthen talking to the crew. Their voices sounded far away, like a radio playing in another room. The inner lining of his dry suit stuck to his body. His teeth were locked in a stiff, tendon-tightened clench. He was feeling that old tightness in his jaw.

Your jaw is too damned tight, he was thinking. You know what that means. Admit it. You want out of here. You do. Can you do any more good here like this? No. It's time to stop everything. It's time to fly away from here. Pull the collective all the way back and get us out. Come on. Pull it back.

He heard Molthen and Sansone talking through the intercom.

"Maybe we should throw out the raft, Rich."

"That's not a good idea, sir."

"Why not?"

"What if *we* go down?"

"Yeah, you're right."

"We're done."

"Maybe we could go down only as low as a hundred feet and try again."

"How? The hoist's birdcaged."

"Yeah," Molthen said. "Well, we got to do something." He looked over at Adickes, but Adickes didn't look back. "You know, there's no way they're going to send another plane out here."

"Sir," Sansone said. "Hold on a second." He had just heard something on the high-frequency radio—a voice.

"*. . . Coast Guard . . .*"

He turned up the volume to the highest setting.

"*. . . 6018 . . . this is Les Dawson of Alaska Airlines, Flight 196 . . . are you out there?*"

Sansone yelped, "Mr. Adickes! Alaska Airlines is calling us on the HF!"

When he heard the words *Alaska Airlines*, Adickes turned his head. It was like he'd come back from a long way away.

"Well, shit, Rich," he said. "*Talk* to them."

Sansone gave his name and the helicopter's number and relayed their latest GPS position. "We've spotted four to five survivors in seventy-foot seas," he said excitedly. "At this time we are making an approach to the water. Do you read?"

"*Say again, 6018 . . . Understand you've crashed in seventy-foot seas?*"

"Negative," Sansone said. "Negative—"

He heard a crackling silence come back, and then: "*. . . please repeat . . . Understand you are in the water? . . . You are egressing into the water?*"

"No!" Sansone shouted. "No! We're *fine!* We're circling the survivors at three hundred fifty feet! We've been trying to hoist them for the past hour with no joy!"

He waited for an answer. If word that they had crashed got back to the air station . . .

"*Sorry about that,*" the pilot said. "*Now understand you are okay, correct?*"

"Yes! Yes!"

"*Good. Good. Coast Guard 6018, be advised that another aircraft is on its way from Sitka . . . repeat . . . a second helicopter is on its way from Sitka . . . on scene within fifteen minutes . . . over.*"

It's a miracle, Sansone thought. We actually got through to somebody. He thanked the pilot and signed off.

"Mr. Adickes, we got through! We got—"

"I heard, Rich. I heard. Good job."

Adickes had his mind on the fuel gauge. It was well below a half tank. If the fuel-burn calculations you made back in Sitka Sound were accurate, he thought, we've got only ten minutes of on-scene fuel left. If we stay here any longer, we won't have enough to make it back. And how can you be sure you allotted enough gas to fly back against these headwinds?

"Okay," he said over the intercom. "Listen up." He paused. His voice sounded strange to him. Tinny. "I think we've done everything that we can do out here. And now we know that the 6029 is only fifteen minutes away, so, uh, it's—it's time to go."

He got no response.

"We can't do any more good out here," he went on. "We're almost at BINGO fuel and this weather is as shitty as it gets. We don't need to make ourselves into another SAR case."

Still no one responded.

They know, Adickes said to himself. They know what I'm saying is crap. But I've got to say it. I've got to say it and we've got to leave four or five guys in the water. We didn't get them. They're going to die out there. We're leaving them to die. Live with that, Adickes.

"I'm not sure anyone's going to get those guys in this shit," he said.

"And besides, we can do more good back at the station helping out the other crews than continuing here."

Well, that was it. No one had spoken up or even tried to interrupt him. They know, he thought. Now they got to swallow it.

He sighed.

"Rescue checklist, part three."

The Jayhawk rose in a slow, jerking, climbing stagger and, shuddering against the wind, started for Sitka. Molthen looked out his side window. Four hundred feet below the ocean was a series of white, scalloped crests. The strobe was gone.

He looked over at Adickes.

"Bill?"

"Yeah?"

"I'll take the controls."

Adickes nodded.

"You all right?"

"Yeah."

"You did good, Bill. You did real good."

Adickes eased his grip on the collective and the cyclic, and pulled his feet off the pedals. He would have to keep a close eye on their heading. Now they would race the fuel gauge to the air station.

"Sure," he said. "We did great. Just do me a favor and not talk about it."

F O R T Y

You were lucky, David Durham thought, that the 6018 didn't go in the water. That would have made things complicated. No, he thought, hellish. Thank God that didn't happen. They aborted and left four, maybe five survivors in the water. But at least they didn't crash. I hope they've got enough gas to get back. Maybe they'll ditch onshore. Well, it's up to us now. If only this storm would lay down a little bit.

"Dave?"

He turned his head. Russ Zullick was pointing to a tiny screen on the dash panel, the terminal collision avoidance detector. There was a blip on it.

"I think that's them," Zullick said. "That's the 6018 on the T-Cast. They're coming right at us."

"How far are they?"

"Thirty miles, but closing fast."

"Let's try them again on the radio," Durham said. "We've got almost zero visibility. I do *not* want fly into those guys. We've got to talk to them. Right now."

"Okay."

There was heavy static but once the distance between the helicopters narrowed to within twenty miles Zullick was able to talk freely with Bill Adickes. They agreed to fly at different altitudes. The 6018 would climb to an altitude of seven hundred feet; the 6029 would stay at three hundred feet. The chances of a head-on collision were slim. But this wasn't a night for carelessness.

"What are conditions on scene?" Durham asked.

There was a pause. *"Listen, you guys,"* came Adickes's voice. *"These are* extreme *conditions. I can't stress that enough. It's like nothing you've ever seen before."*

A sharp, prickly wave spread over Zullick's back and arms. "Okay, Bill," he said. "We understand that. What are the wind speeds?"

He heard only static.

"Say again? Wind speeds?"

It sounded as though Adickes had said sixteen to seventeen knots, but Durham shook his head when he heard the figure. "No way that's right," he said.

"He must be saying sixty or seventy," Zullick said. Raising his voice, he said into the microphone: "Bill. Please repeat wind speeds and provide wave heights on scene."

There was too much interference. To Zullick, it sounded as though Adickes was saying wind velocities were in excess of seventeen knots, with fifteen- to seventeen-foot seas. But judging by the tailwind, Adickes had to be telling them to expect seventy-five-knot winds, 110-knot gusts and seventy-foot waves, or higher.

Zullick looked at Durham.

"Not too nice," Durham said.

"No," Zullick said. "I'd say we are *in* for it tonight."

Into the radio, Zullick said: "We read you on the wind speeds and wave heights . . . Please give us the survivors' latest position . . ." He listened and jotted down the coordinates on his kneeboard. "Roger that. Understood."

The transmission was breaking up again.

". . . *rogue waves . . . downdrafts. You . . . expect them . . . don't get too low. Don't . . . hovering below a hundred feet. We tried . . . couldn't get . . . basket will . . . searchlight pretty much ineffective. Use lots of flares . . . may lose comms . . . where . . . C-130 . . ."*

Zullick turned down the volume.

"Well," Durham said, "that's all we get."

"Yeah," Zullick said, "but I did get the survivors' position. I'm going to put it into the flight computer." He tapped in the coordinates that were on his notepad.

"Okay, it's in," he said.

"Nice," Durham said. "Now let's get there."

With the exact coordinates, there was no need to leave on the finder, so Zullick switched off the homing device and set a course for the EPIRB. Hurtling along at a 225-knot-per-hour clip, they closed in on the position less than ten minutes later.

"Moving up on it," Zullick said. "We should be over them . . . wait . . . wait . . . *now!*"

Durham flipped on the flood hover lights, which threw cones from the belly of the helicopter, and aimed the Night Sun straight down so that it would not light up the precipitation engulfing their windscreen and blind them. Zullick peered out through his spotter's window.

He could barely see the ocean, even with the night-vision goggles down. He looked and looked. But he didn't see a strobe. He didn't see a vessel. He didn't see anything but the tops of swells whitened by wind.

"There's nothing there," Zullick murmured incredulously. "Where did they go?"

Three hundred feet below and ten miles downwind, the survivors were drifting fast. The rope tying them together had come loose. All they had left from the *La Conte* was one buoy ball, tied around Mike DeCapua's waist, and the handheld 406 EPIRB.

They were still bobbing in pairs, Bob Doyle supporting his skipper, Mark Morley, and Gig Mork floating on his back and holding Mike DeCapua up on his chest. It was so dark they could barely make out one another's form, except when the flash of the beacon painted everything around it a skeletal white. The combers were barreling down on them with no warning and burying them for a half minute at a time.

Each time they went under Bob Doyle could feel the sea tear another layer of heat off his face. One thing he had always dreaded was the cold. He could stand cold as well as most any man, unless it went for too long and wore him down, but now the cold was so numbing that he was beginning to worry about hypothermia and possibly losing a limb. The hope that had electrified him upon seeing the searchlight in the sky, and the anger and emptiness he had felt watching the lights of the first helicopter fading and then vanishing, had passed. They had been replaced by worry. He was worried for his own well-being, certainly, but he was even more worried about his skipper, who he suspected had started his final slide into hypothermia. Morley's suit was torn at the leg. He had been taking on thirty-eight-degree water since the boat sank. That was more than four hours earlier.

The way he is drifting in and out of consciousness, the way he is talking, the way he is shaking, Mark's core temperature must have dropped below ninety degrees, Bob Doyle thought. It could be as low as eighty-eight. But how low can it go before he suffers a heart attack? I guess it depends on the individual's body fat. Mark's got plenty of that, thank God. And he is still breathing. He's breathing and that's all that matters. I've got to keep his head up. He doesn't even have the strength to keep his head from falling forward into the water. I've got to watch that. If his head goes under for more than ten or twenty seconds, he's had it. That's all it will take. I can't let him drown in my arms. I can't let that happen.

"Keep your head back," he said into the skipper's ear. Morley had just snapped into consciousness again.

"Where are we?"

"I'm right with you, buddy. I'm right here."

"They left us," Morley said. His speech was slurring. "I know it . . . left us to die."

"No," Bob Doyle said gently. "No. We're not going to die. They know we're here. They know where we are. They're tough. They'll be back."

"Why?"

"Why what, Mark?"

"Why, Bob?"

It was a whisper of a voice now, bled of any emotion.

"Hang on," Bob Doyle told him. "Remember what I told you?" He got no answer. "Remember, Mark? Remember how I told you that we're going to put you in the first rescue basket that gets close? You're going up first, man. First man up. Ask Giggy, here. Giggy'll tell you."

"I . . . I know . . ."

"First one up—that's you. I promise."

"I can't take it anymore."

"They'll drop the basket near us and I'll get you right in." A wave crunched them down. They came up and Bob Doyle went on, "You won't have to worry about a thing. Hear me? I'll put you right in the basket. They'll haul it up. They'll get you into a thermal bag. You'll be good as new."

A comber caught them from behind. When the sea spit them back out, Bob Doyle sputtered and called out his skipper's name. But there was no answer.

"*Mark!*"

Just then Morley burst through the surface. Bob Doyle grabbed him and shook his head.

"Hey! Wake up! Wake up!"

"Oh . . . oh . . ."

"Don't do to that to me, Mark," he said. "Stay with me, Mark! Stay with me!"

"I can't . . . I can't keep my eyes open . . ."

"Hey. *Hey!*"

Morley fell limp in his arms.

"Mark!"

Another wave rumbled down on them and the next thing Bob Doyle knew, Mork and DeCapua were floating alongside him.

"How's Mark doing, Bob?"

"Not good."

"Where the fuck are those Coasties?"

"How the hell do I know?" Bob Doyle said. "They ran out of gas. Maybe they went back to get more."

"Back to Sitka?"

"Who knows?"

"They ain't coming back."

Bob Doyle coughed, rackingly. His mouth and throat were on fire.

"You quitting on us, Gig?"

"Who said I'm quitting?"

"Well, they won't quit either. They'll be back. I know it."

"Yeah?"

"Yeah."

Mork had DeCapua by the armpits and was leaning back, floating. DeCapua had not said anything for a while. He was groaning.

"Listen, you bastard," Mork snapped at him. "You got to help yourself. You know? I'm getting tired. I'm getting real tired. You got to do more to help yourself."

"Sorry, Giggy."

"Fuck you're sorry."

"I can't see."

"Oh yeah? Fuck you."

Another crasher hit, and this time Bob Doyle lost his hold on Morley. He felt as if he were suspended in thin air. He felt his father's strong hand, shoving him backward, out of their aluminum skiff . . .

He broke the surface.

"You son of a bitch!"

It was Mork's voice. That much he knew. He was screaming. Bob Doyle wiped his eyes.

"Get back here!"

Where was Mark? Where? In a panic, Bob Doyle whirled around, thrashing the water, kicking, feeling about him, and then Morley surfaced. Bob Doyle lunged for him, seized him by the hand and hoisted him up on his chest. He put his ear to Morley's mouth. The skipper was still breathing.

He heard Mork hollering.

"*I said get back here!*"

Something was missing. At first, Bob Doyle did not know what it was. He could not see his hand in front of his face. Everything had turned black. He bobbed, holding fast to Morley, swiveling his head, trying to see what was different. And then it came to him. The strobe was gone.

He thought: Mike's got the EPIRB.

Where had he gone?

Mike DeCapua was in the fight of his life now. He was fighting the darkness, fighting the snarling, black hills of water, fighting his panic. Waves were blindsiding him, exploding in his face. He heard voices. They came from far off, then closer, then far away again. His stomach hurt. His lungs hurt. Everything hurt. He was flailing his arms, wriggling his hips, telling his legs to kick. The legs were not obeying.

He cried out for the others. A wave answered him.

When he came up for air, he clawed at the water and then a white flash blinded him. He fell on his back and a wave slipped under him, lifted him high, and then he felt himself sliding down into a trough.

A flash went off in his face again.

I've got the EPIRB, he said to himself. I ain't lost. They'll see it. They can see the strobe.

He heard the voices again.

"*Mike!*"

That was Bob Doyle's voice. He was sure of it.

Then: "*Get your ass back here, you lousy son of a bitch!*"

That's Giggy, he thought.

"Mike!" he heard Bob Doyle shout. "We're right over here! Come over this way!"

It was hard to tell how far away they were; hair was plastered all over his face. Not even the wind would blow it off.

He shouted:

"Where are you guys?"

"We're over here! Right over here!"

"I can't do it!"

"Get your ass over here! We can't break up to go get you! Get over here, now!"

"I can't!"

"You got to get over here, Mike!" That sounded like Bob Doyle again. "I can't come for you. I'm holding up Mark! You can do it!"

He heard Mork again.

"You want to kill yourself? Go ahead! I don't give a fuck. But throw us the fucking strobe! We need that strobe!"

He started slowly, banging the surface with his heavy, aching arms. His legs were not working well. The muscles were stiffening from the cold. He heard Bob Doyle and Mork shouting. He swam toward the shouts. He could feel the current swirl him and he splashed on until he thought his arms would drop off from the pain. He felt himself ride the crest of a swell, then tumbled down into a trough, and still he kept fighting his way, slugging at the ocean enveloping him.

"Move your ass!"

"Keep coming, Mike!"

The spray lit up in the blink of the strobe like a sparkler throwing burning sparks. There were walls of water toppling all around him. He heard them crashing even after he'd gone under.

He came up spitting and swinging his arms and he hit Mork in the face.

"Whoa," Mork shouted. "It's me."

"Help," Mike DeCapua said.

"I got you."

"Help."

"I got you," Mork told him. He pulled DeCapua up on his chest and lay back and rode a swell. Mork said, "Say, why don't you give me that strobe? I'll hold on to it."

"Like fuck you will," DeCapua told him.

In the back of the helicopter, Sean Witherspoon was sitting with his eyes tightly shut. He had his gunner's belt on. He was leaning against the back wall, trying not to make any sudden moves.

The seat went sailing off so he sat up sharply and looked ahead at the cockpit through the indigo light to make it stop. It's the vertigo, he thought. Don't look down. He felt around under his seat for the plastic bag he'd kept his flight helmet in. He found it and opened it. His toes were gone. He couldn't feel them anymore. His mouth tasted of bile. Keep your head back. You've been here before. Keep your head back and try to relax.

The cabin was spinning fast. But he held it. He held it and held it and held it but then suddenly he couldn't fight it anymore and hot vomit was coming up. It was all he could do to keep it out of his own lap.

His dry suit stuck to him. The cabin was going round and round and round. He grabbed Rich Sansone's right arm. Sansone was on the radio, keeping their guard with Juneau.

"Rich," Witherspoon said. "Look at my hands."

"They're shaking."

"I'm cold. I can't stop it. I'm shaking real bad. I can't stop shaking, man."

"What's wrong?"

Witherspoon blinked and then his eyes closed and he threw up on Sansone's leg. The rescue swimmer turned on the intercom.

"Hey, Mr. Adickes," he said. "You're going to have to take over comms for me."

"Why?"

"We got a problem. Sean's not doing great."

"What do you mean Sean's not doing great?" Bill Adickes sounded incredulous.

"He can't stop shaking," Sansone said. "He's shaking out of control. He's throwing up on me."

"What?"

"He's going into convulsions."

"What?"

"Convulsions."

"Okay, okay. I got the radio guard. Go ahead. You take care of Sean."

"Roger."

Sansone laid the big man out on the cabin floor, pulled out a thermal sack—a hypothermia bag—and slipped it over him. Sansone did not want to cut the dry suit. If we have to ditch, he'll freeze to death in a New York minute. Who am I kidding? We all will.

He slipped a mask over Witherspoon's mouth and opened the valve on the oxygen tank.

"Do you know where you are?" Sansone asked him gently.

"No."

"What's your name?"

Witherspoon just shook his head.

"How are you feeling?"

"I—I can't stop shaking."

"Are you sure?"

"Help me."

"Everything will be all right. Everything's going to be fine."

Witherspoon turned his head and threw up again. Sansone rolled him over on his side and held a plastic bag near his mouth.

"Sean?"

Witherspoon had gone pale; fat beads of sweat had broken out all over his face and neck. Sansone put the water bottle to his lips but he wouldn't drink. His eyes fluttered open and closed. His legs jerked. The flight mechanic's respirations were well above twenty, his pulse was twice what it should have been and his core temperature was 92.6 degrees.

"Mr. Adickes?"

"What now, Rich?"

"How far are we from base?"

"I don't know, Rich. Another hour—maybe an hour and a half. Depends on the winds."

"Sir," Sansone said. "Make sure there's an ambulance on deck when we land."

"What for?"

Sansone said, "I think Sean is going into shock."

For twenty minutes the crew of Rescue 6029 searched the seascape for the strobe. They saw nothing but froth. The waves were running everywhere. In the eye of the storm there was no pattern to them or to the wind and the seas stacked up jaggedly, in white peaks. The turbulence was a lot worse, too. It felt like flying in the mountains. The waves were almost like mountains and the wind would come screaming up the back sides of the waves just as it would howl up the side of a glacier, and many times it caught the helicopter off guard, always from a different angle, and flung it backward into the wind stream or down into a trough. The wind was confused and the sea was confused and the crew was confused and startled by the ferocity of the storm, but they kept combing the cratered ocean in three-mile-long search legs, hoping to spot the flash of a beacon the size of a bowling pin.

When he was not monitoring their altitude and fuel-burn rate or backing up David Durham on the controls, Russ Zullick kept scanning the ocean with his night-vision goggles.

We're wasting too much gas, he was thinking. But there is nothing else we can do except this. And hope. Sure. We could use some of that.

He snapped on his direction-finding radio again and maxed out the volume so as to catch even the weakest signal from the EPIRB. He remembered what an old instructor in flight training school once said: *Fishermen tie themselves together when they go in the water and they tie the EPIRB onto themselves. If they go down, they all go down together.*

Zullick spoke up. "Whoever it is, they drifted far."

Durham did not answer him.

All of his energy was going into monitoring his dash instruments and his cyclic and collective inputs. Zullick noticed the sweat sliding down his cheek.

"It doesn't make sense," Zullick said. "I mean, there was only, what, fifteen or twenty minutes lag between when the 6018 left and when we got out here. How far out can these guys be?"

"I'm going to make a turn," Durham said. "Back me up on the controls."

"Roger that."

Zullick scanned the display. "Let's try looking north and west of here."

Earlier, he had set his BINGO fuel mark at 2,800 pounds, which was about five hundred gallons of gas. The H-60, under normal conditions, burned about 1,200 pounds an hour. Zullick had figured they would use one-eighth of a tank to get on scene and another quarter of a tank to perform the rescue. That left a half tank of fuel to fly home with. But he had not counted on 75-knot headwinds and 140-knot-plus gusts. Just attempting a hover was degrading their fuel endurance. And they had already been airborne an hour and fifty minutes. They should have been well into hoisting by now, not searching.

Zullick was thinking about his reserve and recalculating their fuel-burn rate on the computer when he heard Chris Windnagle, the flight mechanic, shout over the intercom: "I see a light! I see a light!"

"Where?"

"There!"

"Where? Where?"

"Turn us around! Ninety degrees! Hard right!"

Doing that, however, was not possible; they were cocked hard into the wind line. If a gust snagged their nose they would flip over like tumbleweed.

"I'm going to turn her easy," Durham said. "Back me up."

"Roger," Zullick said.

But when Durham tried to swing them around, the helicopter just kept on sailing downwind, as if skidding on ice, until they were a mile downwind of the survivors.

"Son of a bitch," Windnagle said. He had his night-vision goggles down and was looking out his spotter's window. "I think I lost it."

"We'll find it," Zullick said.

"I *lost* it."

"Relax, Chris. We'll find it."

To get back, Durham tried crabbing the helicopter—swinging its nose and tail back and forth like a scampering crustacean. It took another ten minutes just to return to the area where Windnagle had spotted the strobe. They all peered out.

There was no strobe, no survivors, no telling the difference between a wave and trough. Zullick checked his watch: they were now just a minute shy of the two-hour flight mark. It was almost time to return to base.

He turned to Durham and said, "Almost two hours even, Bull."

Durham nodded.

"Wait a minute," Zullick said. "You hear that?"

"I hear it."

A pinging noise was coming through their headsets. That was the sound the direction finder made when locked on an EPIRB signal. The pinging was faint. But it was there, all right.

"I got it, I got the position," Zullick said, punching the coordinates into the computer.

"Okay, let's do it."

Putting the wind just off their nose, angling them in the direction the needle was pointing, Dave Durham stepped up his airspeed to seventy knots. The helicopter seemed to slide backward. He fed the engine more gas. The aircraft did not budge.

"Well."

He pulled more power. Now the airspeed indicator showed eighty-five knots.

"Come on," he said.

The helicopter nudged forward. It was strange to hear and feel the engines working so hard and to be making three knots over the ground.

"That's the stuff," Zullick told him. "Let's move up on it nice and slow."

Nose shuddering, turbines whining, direction finder pinging not so faintly now, they crept upwind. The needle had stopped swinging back and forth. It was pegged at eleven o'clock. Zullick flipped down his goggles.

"Come left," he said, and Durham did, until the finder needle was pointing dead ahead.

"All right," Zullick said. "Stop your turn. Fly this heading."

Durham stepped them up to ninety knots of airspeed. The helicopter moved ahead sluggishly.

"Take us ahead just a touch," Zullick said.

Gusts batted them from both sides but Durham did not break the rhythm of his movement on the pedals and sticks. Suddenly, as if a curtain had been yanked away from the windscreen, Zullick spotted a green flash in the blackness.

"I see a light!"

It was a bright flash, like a mirror turned in the sun. Another fifteen seconds passed. He saw it again. The blinking light was moving away from them in the great, foaming expanse that was the sea.

"We're not losing it this time, guys," Zullick said. Then to Durham: "Bull, step us down a little closer."

Typically, pilots pulled the airspeed off while simultaneously reducing their altitude. But with such winds they had to fly one direction at a time— slowing first to a halt, then descending, then picking up speed again.

As they dropped down, it was as if a camera lens were being brought into focus: Zullick now saw the blinding flash of the strobe a hundred yards off their nose at two o'clock. Through the chin bubble he saw moving pieces of reflective tape—the arms of survival suits, waving.

"There's people down there," he said. "And they're alive."

Once the Alaska Airlines jetliner began circling the Fairweather Grounds, Ted LeFeuvre heard the radio chatter start and his heart seemed to start again with it. Listening to the radio, he and Guy Pearce had just heard that Rescue 6018 was still airborne. Its crew was aborting its mission. But the second helicopter, the 6029, was arriving on scene.

Ted LeFeuvre let out a heavy sigh.

"We've got to get a third helicopter out there to protect that second crew," he said.

"Absolutely."

"Put another crew together."

"Right away, sir."

While Pearce worked the phones, Ted LeFeuvre studied the ocean charts and thought out his next move.

So now you've got people in the water, he thought. Four to five people, by the sounds of things. If they're still alive, those people are cold. Well, more than cold. They've been in the water three and a half hours now. *In thirty-eight-degree water.* Most guys, even a fat one wearing a Gumby suit, will freeze to death in three to five hours in that kind of water. Okay, maybe six hours. Max. After that, you're talking miracles. That means if we don't get those people in the next two and a half hours they will die. They will die. That is clear. Too clear. But that isn't a license to get reckless. You didn't put those fishermen out there. You have two aircrews to think about. And soon, a third crew. Keep them safe. What good is it if we race out there and make mistakes and kill our own? No good. *There are no old, bold pilots.* Remember that from flight school? You can't take risks over and over again and survive. There is a balance and you have to analyze the risk. You analyze it. Then you take it.

Those fishermen took it. They didn't analyze well. But that's their problem. Your problem is getting a third crew out there and keeping them alive. And not to worry. You cannot worry or show nervousness or fear. If you give yourself that luxury the fear will spread and infect everyone around you. When people are afraid, the system breaks down. And then we really *are* in trouble. So far you've behaved all right. But keep your cool. Compartmentalize. You were always good at that.

And what about that business of seventy-foot seas? That's got to be an exaggeration, he thought. Those would be *huge* seas. It's unlikely they're that big, even if the winds are, as they reported, seventy knots sustained. But don't discount everything they tell you. When you start shutting out information you are asking for it. Still, seventy-foot seas? No, I don't think that can be right. And why didn't the meteorologists pick up on an approaching hurricane? There was nothing in the forecast about something like this. Then again, he thought, when was the last time you believed a forecaster?

"Captain?"

"Yes, Yogi?"

Pearce turned to him and said, "I've got the crew."

"Good."

"It's going to be Lieutenant Steve Torpey, Petty Officer Fred Kalt, and Petty Officer Mike Fish." He paused. "And you."

"Me?"

"Yes, sir."

Ted LeFeuvre stood there, feeling his heart slowly sink, feeling the hollowing dampness work its way into the pit of his stomach, feeling the cold wave of dread sweep over his back and neck and shoulders.

You know what it's like to fly in a hurricane, he thought. You flew in Hurricane Juan. That was more than ten years ago in the Gulf of Mexico. You remember. That was forty-foot seas and sixty-knot winds. You remember how difficult that was? And that was during the day. And you were thirty-six. You're a CO now. You're supposed to write the awards for these guys. You're not supposed to be out there in that mess. How can I bow out of this gracefully?

He said: "Yogi, you know, I think you'd be a lot more help to Steve Torpey in the cockpit than I would be."

"Uh, Captain," Pearce said, "I won't be able to go until at least four o'clock."

"Why?"

"Well, sir, I was over at the Eagle's Nest tonight. Almost all of the pilots were there. I'd already had a couple of beers when Mr. Durham called."

"Oh," Ted LeFeuvre said quietly.

"Mr. Durham asked if I'd come in to work the desk. So I did." He gave his captain a sideways glance.

"Sir," he added, "there is no one else."

Well, that's it, Ted LeFeuvre thought. You're too old for this. You are entirely too old for this. But you heard him. You're the last pilot left. Talk about scraping the bottom of the barrel. Go on. Quit stalling, you gutless wonder. Tell him you hoped it would be like this. Say something cute.

But all that slipped out of Ted LeFeuvre's mouth was a short, stiff "Okay."

F O R T Y - T W O

By eleven that night the air station was at full tilt. People had been coming in all evening, even those who were supposed to have the weekend off. There were more people working that night than on any day that winter.

Mechanics were towing helicopters back and forth; flight engineers were loading flares, gear and fuel; aviation techs and pilots were poring over aircraft maintenance records, inspecting engines and rotor heads; electronics crews were running last-minute circuitry and instrument checks; rescue swimmers were testing immersion suits, wet suits, life vests and mesh vests; supply clerks and petty officers were staffing the phones, working radios, plotting positions on aeronautical charts and gathering the latest weather information; cooks were rushing out eggs, pancakes, bagels, sausages, hot soup, sandwiches and coffee; and even the airbase doctor reported for duty. It was the first time in recent memory that the light in his office had been on past five o'clock.

Ted LeFeuvre was up in his office, changing into his flight gear. He was thoughtful and deliberate, not plodding, but not rushing himself. Everything had been done that had to be done up to that point. All orders had been given. Everyone knew exactly what he or she was to do that night and in the morning. He was thinking either the rescue would turn out well with the coming of daylight or it would not. I believe it will it turn out according to the Lord's will. That is all we can do, anyway. Believe.

And then, for the briefest of moments, he had a creepy, unreal feeling about everything that was going on, as if he had been through this rescue once before.

And maybe you have, he thought. Remember New Orleans? Ah. Nice try. This is not the Gulf of Mexico. This is more. Much more.

But there was no panic or excitement in his heart when he admitted this to himself. Those feelings had begun to drain out of him the moment that he had accepted in his heart that this mission was the Lord's will and that God was leading him toward something. Perhaps he was a little jittery at first. But as soon as he saw Steve Torpey stride into the operations center and they began debriefing and drawing up a flight plan, the nervousness dripped out of him and he felt the calmness of old return. Then he knew he was all right.

Now, as he pulled on his dry suit, there was nothing but the calmness. His mind was clear and he was thinking the way he had always liked to think. Analyzing each facet of the mission. Ordering his priorities. Excising all nonessentials. Uncertainty, fear, loneliness—he had to compartmentalize them.

It was strange, he thought, that he had once been incapable of saving his own life, and yet here he was, ready to step into a helicopter to save someone else's. Well, this is God's will, he thought. Only He knows why this is important. Just go with the flow. That's all you can do.

He was sure now that the third and last helicopter on the base, the 6011, would take off with him flying it. That was certain. What else? He was flying with the Coast Guard's best. These were men who flew helicopters not because they wanted medals or thrills on a Friday night. They were going out because they were professionals and because others depended on them to act like professionals. Well, maybe Torpey still had a little top gun in him. That was to be expected. He was young. He still had the zealousness. But he also had the gift. Oh, did he have the gift. He just hadn't been tested yet. Neither had the others. Well, Steve Torpey and Fred Kalt and Mike Fish. Prepare to be tested.

The mission would end. It could end badly. But it would end, and the Lord would be watching. He was sure of this and the thought comforted him. Certainty was always comforting.

He stood there, suited up now, feeling the tight, polar-fleece dry suit warming his skin, the dragging on his neck from the pull of the survival gear, the weight of the scuba tanks on his back, and the puffiness of his inflatable life vest. In his left hand he held his flight gloves and in his right, his flight helmet. It was an old helmet and had a few dings on it, a few scrapes. It was well worn but it had served him well for many years and had always stayed dry and out of the water.

Ted LeFeuvre turned off the light and pulled the door until it clicked and started down the hall. He needed to go to maintenance control. There was just one more thing he needed to do before getting on the aircraft.

Now the four of them were standing in a circle in the maintenance control center, facing one another.

"That's it. That's the latest," Steve Torpey said. He had just finished briefing everyone on the most recent information radioed in by Bill Adickes, whose crew was just outside of Sitka Sound.

"Okay," said Fred Kalt. He had been anxious to get out on the hangar and check over the aircraft. "Let's go."

"No," Torpey said. He looked over at Ted LeFeuvre. "I want to wait."

Torpey felt awkward when he said it, as awkward as a kid saying good-bye to his girl on the front porch of her house the day before he was to go off to college. It was not usual to hold up a launch. On a SAR case you had thirty minutes to get airborne. If you didn't launch within that thirty-minute window there had to be a darned good reason, and you would be expected to give that reason later in writing.

What made it tougher was that his crew was ready to go. And it was 11:35 P.M.; the survivors, if they still were alive, had been in the water four hours and forty-five minutes. The chances of those survivors drowning or dying from hypothermia were rising by the minute.

He looked at Ted LeFeuvre.

"Why wait?" Kalt said. "Let's get going."

"Hold on, Fred," Ted LeFeuvre said to Kalt. "What is it that you're thinking, Steve?"

"Well," Torpey said, "there are no more night-vision goggles in the station. The other crews took them all. I say we wait for the first crew to come back before we launch. Adickes will be back in fifteen minutes. We can get their goggles, debrief and launch."

He looked around.

"I'd feel a whole lot better flying at night with the goggles."

"Agreed," Ted LeFeuvre said. "We wait."

The others nodded.

"Anything else?" Ted LeFeuvre asked.

"How about more flares?" Mike Fish said.

"Where are they?"

"There are crates of them in the pyrotechnics locker." It was a small depot on the opposite side of the base. It would take at least ten minutes to drive over and get the flares.

"Well," Ted LeFeuvre said, "go get them. Let's take every flare we can get our hands on."

"Yes, sir," Fish said, and he darted off.

"Sir," Kalt said, "I think it might help if we dressed the rescue basket up with chem lights." Chem lights were pencil-size, plastic tubes filled with chemicals that glowed a phosphorescent green when they were snapped in the middle. "There'll be no way for them not to see it."

"Good idea," Torpey said.

"You know," Ted LeFeuvre said, "one of the biggest problems we had when we flew in Hurricane Juan was getting the basket to the survivors. With sixty-knot winds, that basket went trailing way behind us. It was hard for the flight mechanic to gauge the delivery of the basket because it was paying out at a really severe angle. If we could weigh it down somehow, I know it would help."

"How about some shot bags?" Kalt said. They looked like beanbags, but were filled with lead shot. Each six-inch-square bag weighed fifty pounds. "We could lay them across the bottom of the basket so the cage doesn't get whipped around."

"Great idea, Fred," Ted LeFeuvre said. He stopped. "Fred," he said, "what do you say about us taking along another flight mechanic?"

"No," Kalt said. He looked at Ted LeFeuvre squarely when he said it. "I really don't need anybody else, sir."

Ted LeFeuvre nodded. "Okay," he said.

"Is that it?"

Within minutes Mike Fish had returned with several crates of Mark-58 flares, and he and Fred Kalt loaded them in the aircraft while Torpey and Ted LeFeuvre reviewed the 6011's maintenance records and signed the sheet for the aircraft. Then the two pilots picked up their helmets and extra gear and walked out through the hangar and across the tarmac.

A fine sleet crackled on the concrete. The wind was creaking the pines on the mountains. They walked out to the far corner of the runway. The Jayhawk sat idle on pad number three.

Fred Kalt looked up from his preflight inspection checklist and shook his head when he saw Ted LeFeuvre approaching with his duffel bag.

"Say, Captain," he asked, "what's in the big bag?"

Ted LeFeuvre shook his head. He always brought along the parachute bag and they always razzed him about its size. In it he carried an extra flight suit, clean boxer shorts, T-shirt, long johns, overnight bag, toiletries, helmet sack and day parka.

"Need a hand lifting that, sir?"

"No, thanks, Fred."

"Hey, Mike, give me a hand with the captain's bag."

"I've got it, Fred."

"Now where can I put it?" Kalt was scratching his helmet and feigning a confused look as he stared in the cabin door. "Hey, Mike. Where are we going to stow all of the captain's cargo?"

"Listen, Fred," Ted LeFeuvre said, "I've broken down too many times in the Alaska tundra and nearly frozen to death. This helicopter's big enough to carry my stuff."

Kalt and Fish yucked it up good. Torpey turned away, grinning. Ted LeFeuvre looked at him.

"Did I say something?"

"No, sir. No."

Ted LeFeuvre walked over to the right side of the aircraft, opened the cockpit door and climbed in. I know I can fly this thing, he was thinking. I'm shaky on that computer. But the aircraft I know I can fly.

Steve Torpey stood at the door, watching him strap in.

"Say, Captain?"

Ted LeFeuvre hadn't pulled his helmet on yet. He was struggling a bit with the seat buckle.

"Captain?" Torpey asked. He stood there awkwardly. "How do you feel about the right seat?"

Ted LeFeuvre looked up.

"What I mean is—" Torpey paused. "What I mean is, sir, uh, how do you feel about the right seat?"

It was not every day that a lieutenant asked an airbase commander to slide over to the left seat, the copilot's seat. Ted LeFeuvre looked at his young pilot, looked down at his seat buckle and nodded.

"Steve, I know exactly what you are saying," he said. "I'd probably ask the same thing if I were you."

And without another word, Ted LeFeuvre stepped out of the cockpit, walked around the front and, opening the opposite door, climbed into the copilot's seat. Torpey watched him, and then took his seat.

They started the preflight checklist. As he was about to reach up to hit the starter button, Ted LeFeuvre said, "Steve?"

"Yes, sir?"

"There's one thing you should know before we take off."

"What's that, Captain?"

Ted LeFeuvre grimaced. "It's just that, well, I'm not very good at navigating with the computer."

It was too dark to fly close to the water. Sky and swells had merged into one black mass.

"Let's put some flares out," Dave Durham said to Chris Windnagle. "I'm going to need some visual reference. I can't find the ocean."

"Roger that."

The Mark-25 flares were two feet long and as wide around as a dinner plate. Windnagle cut the heavy metal bindings, tore the Styrofoam casings off three smoke flares, lined them up alongside the jump door and yanked the tabs.

"Flares are ready to go."

"Say when," Durham said.

Just then a gust uppercutted the helicopter. The floor heaved and the flares went tumbling.

"Son of a bitch!"

"What is it?"

Windnagle lunged for the flares, scooped them up and cradled them to his chest. Don't you lose any of these sons-a-bitches, he thought. Not now.

He crawled on his elbows and knees to the edge of the jump door.

"Ready to deploy the flares," he said.

"On my signal, then."

Durham overshot the survivors and banked the helicopter forty-five degrees. As soon as he rolled level, he shouted: "Deploy flares!"

Windnagle pitched them out in two-second intervals. They burst to life upwind of the strobe.

"Flares are in the water!"

Durham turned the aircraft's nose to go downwind; within seconds they were sailing off at 140 knots.

It took them another ten minutes to beat their way back to where they had made the flare drop. When they could see the flares rising and falling on the waves below them, Russ Zullick called for the crew to prepare the cabin for hoisting. A. J. Thompson, the rescue swimmer, hooked up his gunner's belt and crawled over to the edge of the jump door.

He looked down.

Below, the flares shone like bright candles bobbing in a tank of jumping oil.

"Mr. Zullick," he said, "I could get those guys in the basket. I'm ready to dress out."

"No," Zullick interrupted him.

"But I—"

"No, A. J. We're not going to put you out. I don't know if we could get you back."

"He's right, A. J.," Durham said. "Sorry."

Thompson returned to his seat and went back to working the HF radio. He had been ready to go. He had done his job. But now he doubted they would get anybody.

For twenty-five minutes they hoisted straight by the book: Chris Windnagle operated the winch and conned the pilots into a hoisting position; Dave Durham operated the flight controls; Russ Zullick navigated and backed up the flight commander; and A. J. Thompson maintained their radio guard and helped monitor their fuel and altitude.

But the survivors, clustered around the strobe and what looked like a fish float, were drifting so fast that Zullick had to reprogram his instruments every few minutes just to keep up. The wind was coming so hard that their rotor wash was lagging sixty to seventy feet *behind* the helicopter. And each time Windnagle tried to lower the basket, the wind swept it back at a forty-five-degree angle—dangerously close to the tail rotor.

On the first drop, the basket hit water more than two hundred yards from the survivors. Windnagle winched it in and shouted to Dave Durham:

"Forward and right . . . two hundred yards!"

"What you say?"

"I said, '*Forward and right two hundred yards!*'"

"Okay, okay," Durham said. "Sorry, Chris. I just thought I heard you wrong."

Cable was whizzing out as fast as the winch could feed it and the line was so tight it looked as though the winch might get ripped out of the helicopter. When it hit water the swells would scoop it and flick it into a trough, and a breaker would spike it down underwater. They had floats on the basket, though, and within ten seconds the basket would emerge from the froth and the cable would go slack.

Sometimes Windnagle would give up and try to reel it in, but a swell would rise up beneath the basket and lift it faster than the winch could spool the line, leaving enormous loops of slack cable in the ocean.

There's way too much slack out down there, he thought to himself as he hurried to spool the cable. If one of those guys floats inside all of that slack line and a wave bottoms out, that steel cable is going to go taut and sever him in two.

During one evolution, Windnagle glanced down and saw the survivors had slid under the belly of the helicopter and gone far off to the left of the helicopter.

"Forward one-fifty! Left one seventy-five!"

Durham was now bumping the cyclic against the inside of Zullick's thighs. Under normal conditions, that would be a sign that he was over-controlling the aircraft—stirring paint, aviators called it. But he had no choice; he was fighting the aircraft now, no longer finessing it. The nose was lifting twenty degrees, dropping thirty, the tail was slewing and jerking, and all the time Durham was fighting to find a balance, his face graying and tightening.

Windnagle, poised at the jump door, shouted: "Hold your position!"

Durham worked and worked and worked, but he was unable to hold the Jayhawk still for more than a few seconds.

"Jesus," Windnagle said. "Why don't you guys do what I'm telling you to do?"

"We're trying, Chris."

"Let's try it again," Windnagle said. "Go forward seventy-five and right forty!"

Zullick was monitoring their altitude and how the waves were changing the reading on the dial. He had imagined they would set a 130-foot hover for hoisting; but every now and then he would see a flare on a wave crest rise up until it was almost level with his own line of sight, and then settle back down again. The radar altimeter was cycling from 40 feet to 260 feet in a matter of seconds.

"Hey, Russ," Durham hollered to him, "are we going up or down?"

"Don't ask me." Zullick was looking out his windscreen. The bars of white light thrown by the flares were weakening. "We better put out more smokes."

They were down to their last one, a Mark-58. Windnagle armed it, and as soon as they looped back around, he let it fly. It hit a swell and shot white light.

They had three quarters of an hour of light left to work with. No more.

During those forty-five minutes, Dave Durham did what he could to keep the helicopter over the survivors. Russ Zullick watched their fuel reserve shrink.

First he reduced their calculated BINGO limit—the fuel they would need to get back to Sitka—by a hundred pounds, then another hundred pounds, and then another hundred.

"Sir," Thompson said to Zullick. "We can't keep dropping our BINGO fuel like this. Soon we won't even have enough to make to shore to ditch."

"We're dropping it another hundred," Zullick said. Then to Durham: "Dave, hold your nose up. Hold it up."

The helicopter was moving around so wildly that dropping the basket near the survivors was like dropping a clothespin into a milk jug from atop a ten-story building. With each drop, though, Chris Windnagle was honing in on the target; by the time they were forty minutes into the evolution he was consistently putting it into the water no farther than twenty feet from the strobe.

On his last drop, the basket splashed down in a trough thirty feet behind the helicopter.

"Oh, man," Windnagle said excitedly. "I got it, maybe, ten or fifteen feet from them."

He waited and watched, but none of the survivors made a move.

"Son of a bitch," he said. "Oh, *damn* it to hell. They're not going for it. I don't think they can see it."

A comber crashed over the basket and cable raced out. Windnagle kept watching the survivors.

"It's fifteen feet from them," he said. "Shit, they're not moving!"

The aircraft was shuddering fitfully now and lurching from side to side, up and down. But there was enough slack in the line so that the basket did not move much. Windnagle shouted:

"Back us down!"

Durham eased off on the gas, pulling the nose of the helicopter up just a touch as he did this, and a gust caught the nose and threw them back and down. Reflexes taking over, he jammed the cyclic to the instrument panel to drop the nose.

The helicopter's tail leaped up.

"No, no, no!" Windnagle yelled. "I said back us down! Back us down!"

Zullick had been listening to the two of them and trying to keep an eye out for rogue waves. When he thought the immediate horizon seemed free of rogues, he glanced at the radar altimeter. It read sixty-two feet.

Just then it occurred to him that he had lost sight of their last flare.

His eyes swept the ocean.

Where is it?

He looked up.

The flare was atop a crest of an approaching rogue—forty feet above the aircraft.

Just as Durham shrieked, *"WE'RE GOING IN!"* Zullick seized the controls. In one motion he leveled the wings, brought the nose to the horizon and pulled power on the collective with everything he had.

"ITO!" he screamed. *"Instrument Takeoff!"*

"Wait!" Windnagle screamed. *"Wait!"*

They were rocketing skyward.

"I still got the basket out!" Windnagle was shouting. "The basket's still out the cabin door! I've got cable out!"

In the pit of his stomach Zullick felt an icy chill, as though an icicle were dripping there, and in his temples the pounding of hot blood, and knowing full well that he could not make a mistake and had no more time to think, he pulled more power. He had seconds now, seconds to get them up and away from the black wall fast closing. He was trying to be loose but firm on the sticks, holding his breath and trying not to think of anything but the wave; to pull them free of the downdraft that was heavier now than it was when he first saw the rogue closing in.

Now they were starting to get forward airspeed in addition to the winds coming at them. Zullick heard screams: the wind, the turbines, his flight mechanic, Windnagle, who was fighting to bring in the rescue basket now whipping like a snapped kite and thudding along the airframe just feet from the tail rotor.

"SHEAR!" Zullick screeched. "SHEAR! SHEAR! SHEAR!"

"I can get it in!"

"SHEAR THE CABLE IF YOU HAVE TO!"

"I can get it!"

Holding the collective back, feeling the sweat sting his eyes and the spasmodic lurch of the helicopter as it ripped free of the downward smothering draft, Zullick heard it—an erratic, stuttering *thump-thump, thump-thump-thump* through the airframe and then a door slamming and a sudden, heavy quiet.

Windnagle punched the rescue basket. Then he stowed it and the hook and slumped back into his seat.

"We *left* them," he said.

Zullick looked over, and seeing Durham, coldly wet and hollow-eyed, perspiring heavily, he said vaguely, "I have the flight controls."

He leaned forward and checked their altitude. They were up to five hundred feet.

"We left them," Windnagle said again.

"Chris," Zullick said.

"No, no." He let out a big sigh. "We left them."

"Chris, if we stayed any longer we were going to be—"

"We left them," Windnagle said. "And they were still alive."

F O R T Y - F O U R

After the hammering clatter of the rotors faded and the last of the smoke flares had burned out, Bob Doyle closed his eyes and pulled Mark Morley higher up on his chest so that his mouth was against the skipper's ear.

"Mark?"

"I gotta," Morley mumbled, the words slurry, as if he had been heavily sedated, ". . . I gotta get . . ."

"Can you hear me?"

". . . on it . . ."

"The helicopter's gone, buddy," Bob Doyle said into the skipper's ear. "It flew back to get some gas. But there'll be another one. I swear."

". . . got to . . ."

The skipper's head flopped forward.

"Hey," Bob Doyle said. "Hey." He reached around and, clasping Morley's forehead gently and easing it back, pulled the skipper's nose out of the water.

"Mark?"

The combers were coming faster now, no more than ten seconds apart. Bob Doyle never saw them coming. But he knew when they were approaching because his body would suddenly lose its heaviness and begin to rise, as a leaf rises in a breeze just before a downpour, and although he knew he was no lighter he kept rising and hearing the hollow, sickening rush of water and then a wave would explode. When it came down on him everything lighted up in flashes and wheeled around and blacked out, and then the next thing he knew he was coming up through the surface and breathing spray and seeing white and hearing only the air-splitting howl of wind.

Then he would claw through the hilly darkness, eyes on fire, head pulsating, stomach bloated with seawater, to search for his skipper. Each time he found Morley floating, head flopping, arms flung out and loose as strands of kelp, breath coming in quick, whining rasps, he would drag the big man up onto his own chest and lie back and pray that his inflated suit would keep them afloat until the next breaker.

Now he, too, was fast taking on water. It had filled the lower legs of his suit and was halfway up his thighs, and each time a wave rolled him he could feel the water slosh against his crotch, sending a horrific, stabbing shock through his testicles that lighted flashes of light in his eyeballs and left him gasping, shivering from the pain.

Stop shivering, he told himself. Goddammit. Stop and stop and stop and stop. You are *not* going to shiver. You can't. Okay, you can quake a little. Go ahead and quake. But no shivering. You can't shiver.

The cold was all through him now, the heavy cold that had numbed away his feet and ears and snaked through the rest of him like a steel, flexible snake, a snake that slid and curled down his throat and down through his chest and stomach to his ankles. Each time a wave spun him or flipped him, the snake coiled firmer and colder in his chest until it felt as though a slippery, freezing eel was slapping around inside of him. He was becoming afraid of the cold now, even though it had been in him for some time, and he wondered what it would feel like to pull a dry warm blanket over his feet and legs and chest for just a minute. But he knew that this was not possible, that there was nothing to do about the cold but take it and take it and take it, and he tried to take his mind off it by thinking about Morley's suffering. I'm so glad I'm not him, he thought. I'm so glad for that. God, I'm glad I'm not him. The cold is bad in my legs and in my nuts and some now down my back but he lost feeling in his lower body hours ago and he is still fighting it. I wonder how long it takes before you have to amputate a limb or a finger because of hypothermia. Don't think about that, he told himself. Don't think about that. But this guy is dying a creeping death. He's hoping and fighting, but he's dying even as he's hoping. Just look at him fight. He's being frozen alive and still he won't give up.

He shook Morley by the shoulder.

Oh God, he's not shaking anymore. He's beyond the shakes. That means his core temperature is below eighty-five. I think that's right. Well, whatever it is, it's bad. When you stop shaking you cross a line and he's stopped shaking. His heart could stop any moment. Oh, *goddamn* those helicopters. Why couldn't they get a basket near us? Now he's going to have a heart attack. Maybe it's the best thing. I don't know. I can't stand watching him go through this, agonizing like this.

What a lousy, shitty thing to think. What is the matter with you? You've got to save him. You've got to get him into a rescue basket so he can be hoisted up to one of those helos. You're going to have to wake him up to do it. There's no way you're going to be able to push him up into the basket. He'll be too heavy, with all that water in his suit. You'll need him to help you. You'll need to wake him and keep him thinking clearly until he's in the basket.

Christ, this water is cold.

Stop thinking about it, he thought. Think about the helicopter. Didn't that second helicopter come real close to buying it, though? Especially that one time the rogue climbed up the hoist cable and took a swipe at them. I don't know how that wave didn't catch the belly of that thing. And Giggy, wasn't he funny screaming at the helicopter like that? *Send the fucking diver down! Send the fucking diver down!* And you were pretty funny, too. Yelling out your name. *Hey, it's me! Bob Doyle!* Real funny, all right. Hilarious.

"Giggy!"

Out of the darkness he heard Mork answer: "What do you want?"

"How're you holding up?"

"Not good. I'm still stuck with this asshole."

"Hey, Mike!"

Mike DeCapua groaned.

"Mike!"

"What?"

"Give me your lighter."

"What?"

"I said give me your lighter."

"What the fuck for?"

"I want a cigarette."

"You're fucking cracked."

"No," Bob Doyle said, "I just want a cigarette. Giggy, give me a cigarette?"

"Kiss my ass."

A swell lifted them and sent them tumbling into a trough. When he came up for air, Bob Doyle yelled out: "I know one of you bastards took my cigarettes! Who took them?"

He heard nothing.

"Bastards! Who stole my cigs?"

DeCapua shouted back, "I ain't giving you shit."

"How about you, Gig?"

"Get your own cigs."

"Aw, come on, Giggy. Give me a cigarette. Just one lousy cigarette."

Another wave buried them, and then another. When they came up, Mork shouted: "Bob, if we get out of this I'll buy you a fucking carton of Marlboros."

"Luckies," Bob Doyle said, "I want Luckies."

"Bastard," Mork said. "Hey, Bob, where are those goddamn Coasties?"

"They went to cash my retirement check"—Bob Doyle gasped for air—"so we can have beer money once we get back."

When he said it, Morley laughed.

"Hey, man," Bob Doyle said to him. "Welcome back."

The skipper didn't stop laughing. His laughter had an artificial sound to it, an accent of drunkenness, a shrill overtone of idiocy. It was eerie. It made Bob Doyle think of people in smocks behind bars with mad-dog eyes and chains fastened to their wrists.

"Hey, hey," he said, "stop it, Mark. Stop it."

And Morley stopped, just as quickly as he had started.

"Take it easy, man," Bob Doyle told him. "Just take it easy, okay?"

"You know, Bob . . . you need to know it, man . . . I want you to know it."

"What, Mark? Know what?"

"You're a good man."

"Sure, sure. You just hold on, now. I got you."

"You are. You're a good man."

"You just hang in there, Mark. You're going up first. You hear me?"

Morley said, "Listen, Bob. You tell my son. You tell him that I love him."

"Shut up," Bob Doyle said.

"You tell him. You'll tell him, won't you?"

"No," Bob Doyle said. "*You* tell him. Hear me? We're getting you in that basket and—"

Morley went limp.

"*Hey!*"

He shook the skipper.

"*Hey!*"

He could hardly feel his hand in the lobster fingers of the survival mitten, but Bob Doyle curled up a fist, as tight as he could make it, and drove his knuckles into Morley's face.

"Wake up!"

He hit him again.

"*Fight! Fight, you bastard!*"

Crying, socking his skipper, again and again, he suddenly heard Mark Morley groan.

"There," Bob Doyle said, "there, that's better. You wake up. Mark? Mark? Hey, skipper, listen to me. Listen. The helicopter's here. It's *right* behind us."

Groggily, Morley said, "Where?"

"Right behind us," Bob Doyle lied. "Pretty soon they're gonna send the basket down."

This was how the four men talked to one another as the ocean raged and they waited for a light to pierce the swarming darkness.

This sea can look so much like a snake, Bob Doyle was saying to himself. The way it undulates, twists. The way it rolls over and strikes. Maybe a python. It's got the green of a python. Are pythons green or is it cobras? Who knows. In the light of those smoke flares it sure looked green, though. That pale, pretty green. She sure is pretty when she's angry, this sea. Like the spray. Looks just like a storm of fireflies, I swear to God. Millions of them. No, it's more like each wave was a huge sparkler, or fireworks showering a field in Vermont with white sparks. On a summer night. Sure. On July Fourth. Showering the sky with diamond sparks. Diamonds and sapphires and emeralds. That's it. The curling part of the waves—they're green, like fiery emeralds.

Christ, you really *are* losing it, Doyle, he said to himself. Getting all weird thinking. That's the hypothermia, he thought. That's what that is. The cold. It's making you see things, say things. Just pray it doesn't make you start doing things.

He lowered his head and said a prayer.

When he finished he looked up and saw his little one, Katie, sleeping under rumpled, white sheets, her blond hair spread over the pillow.

Go ahead, he thought. Talk to her.

Katie? Katie? I'm in trouble, baby. If you can hear me—I know you're sleeping, but if you can hear me, honey, Daddy is in a lot of trouble. I don't want to scare you, honey. But wake up. Please wake up and tell Mommy that Daddy's in the water.

The girl did not stir.

All right. I'll figure out something, baby. I'll try to figure it out. But just let somebody know Daddy's in trouble, baby. And please hurry, Katie. We're going to die soon.

Then, suddenly, he saw long prongs of light—the beams of headlights—and he was certain he was riding a highway in California. Hills of grass went slipping past, brown grass, the kind that hugs the California hills in the fall, and he knew he was doing a hundred along the San Jose highway. The sky was black. Chipped diamonds, loads of them, lay scattered about everywhere in the sky and he had the radio on and the wind was howling off his side-view mirror and the car racing faster and faster, up and down the hills.

He then saw the wave approaching.

It rose up and over the four of them and came down with a deafening crash and he was twirling in the sea again, wrapped in cold black.

Now it was as if he was looking through a fog. Then the fog parted into shreds of bright blue and he was gazing up into unending sky. He was looking up and there was black soil on both sides of him and at his feet. And then he was staring down at a cemetery. He was looking down at a neatly rectangular pit and he saw his lovely ones, Katie and Brendan, sobbing, their hands caressing the casket, and he knew he was in it but could not understand how he could be looking down on it from above.

Katie's hair was a brilliant, frosted white, tucked partially inside of a beret. Beside Katie stood his mother, and alongside her his kid sister, Sally, who looked no older than nine. Behind them stood his brothers and his father and even his uncle Jim, who looked like he did before the cancer, and every one of them was staring blankly at the casket in the pit. He told them in his mind that he loved them all and how badly he felt for having died on them, and how it had not been his idea to die and that if he could have died any differently he would have.

And then he noticed Rick Koval, standing over his casket.

You bastard, he cried out. *You smug bastard.*

Koval was standing over the grave leering with those pig eyes of his and laughing—that boorish, insidious laugh of his, the laugh he so hated. And then he saw Koval rest one hand on Katie's shoulder and, with the other, give his casket the finger.

———————

His head snapped back.

Bob Doyle knew where he was. He knew he was in the ocean, frigid, yet alive. He heard a comber barrel toward him but he did not worry about it.

You, he said, having taken the wave, spinning now in the blackness, are going to survive. Nobody, least of all that bastard, is going to take Katie away from you. You are going to goddamn fucking survive.

Back at the air station in Sitka, the ground crew had towed the helicopter off the launchpad and over to a fuel pit on the far side of the airstrip.

In the cockpit of Rescue 6011, Steve Torpey was listening to Bill Adickes over high-frequency radio.

"*Steve,*" Adickes said, "*it's nothing like you've ever seen before.*"

The first rescue helicopter was flying across Sitka Sound now and the radio transmission was sharp. Still, Adickes's voice sounded rough.

He said, "*Don't be surprised by needing very extreme inputs on the controls.*"

"Okay," Torpey said.

Adickes said, "*The seas are bad. Real bad. Seventy-foot waves with rogues. Watch out for the rogues.*"

"Right."

"*Don't even think about hovering or hoisting from any lower than a hundred feet. Watch for downdrafts. They drove us down right in front of big waves. And the winds are extreme. They hit you from all sides.*"

"Okay," Torpey said.

"*You've never been in anything like this.*"

"Okay," Torpey said. Ted LeFeuvre was listening to the conversation through his headset. Although he appreciated the information, the roughness in Adickes's voice unsettled him.

"What else can you tell us?" Torpey asked.

"*Take lots of flares. Lots of flares. As many as you can. Get them in the water fast. You'll need them for reference. Otherwise, you won't see the water. You won't see anything. It's all black out there. No light. No light at all.*"

"What else?"

"*Load extra gas. You'll need it. And take another flight mechanic along.*"

"Two flight mechs?"

"It's too much for one guy to handle. We lost Sean early on. Now he's vomiting, hyperventilating. The guy's totally dehydrated. We think he's in shock. Is that ambulance going to meet us?"

"It's on the way," Torpey said.

Ted LeFeuvre hit a switch on his headset so that the crew in the cabin could hear him.

"Say, Fred," he said to Fred Kalt, who was sitting back, strapped in, ready for takeoff, "the first helicopter just recommended that we take another flight mech along with us. We've got Lee Honnold in the hangar."

"I don't *need* another flight mech," Kalt said.

"You don't need one."

"Sir," Kalt said, "I can do this myself. I don't need Lee."

Torpey put Adickes on hold for a second. "Fred," he said, "don't you think it might be rough enough out there to warrant some extra help?"

"Mr. Torpey," Kalt said, "you can't *make it* rough enough for me."

Torpey looked at Ted LeFeuvre, and the captain lifted his shoulders to show he did not know.

Torpey went back to talking to Adickes. "All right, Bill," he said. "Thanks a lot for the warning. Can you give us an idea of what the conditions are looking like out there?"

While he listened Ted LeFeuvre saw the nose light of the 6018 in the sky. Then the Jayhawk's silhouette grew steadily and then its hover floodlights danced along the runway. Out front of maintenance control, an ambulance pulled up.

"Hey," Torpey said, looking out his side door, "there's the ambulance."

The helicopter touched down and taxied to a stop in front of the operations center. The jump door flew open and the paramedics hustled over. They eased Sean Witherspoon down onto a stretcher, put an IV in his arm, an oxygen mask over his mouth, and wheeled him quickly over to the ambulance.

The rear doors swung shut and the ambulance roared off, siren wailing. Fred Kalt sat watching its taillights fade in the downpour. He cleared his throat.

"Captain LeFeuvre," he said. "I was thinking." He paused. "Maybe waiting for Lee isn't such a bad idea."

While Lee Honnold was throwing the last of his gear into a duffel bag and bolting out of the hangar deck and across the tarmac to the waiting helicopter, Bill Adickes was walking through the air station, head doing spins and leg muscles fluttering, down a shiny, gray, fluorescent-lit corridor.

When he reached the locker room he pushed open the door. He was heading to the showers in the back when he saw Rich Sansone.

Sansone was on his knees, doubled over a toilet, retching. Adickes stood watching him for a moment. Then he stepped over and put a hand on Sansone's shoulder. The rescue swimmer did not look up.

He was not done yet.

F O R T Y - F I V E

They were on scene thirty-nine minutes later, in complete darkness.

"It's no good," Ted LeFeuvre said. He flipped up his night-vision goggles. "I can't see a thing out there. What about you?"

"Nothing," Steve Torpey said.

"I'm flipping up my goggles. These things aren't doing me any good."

"Me neither."

Seconds earlier, they had roared directly over the fly-to position Russ Zullick had radioed to them as he and the second helicopter crew flew back to Sitka. Zullick had warned them not to expect the survivors to still be at same position, since the drift was so strong. But they had to start looking somewhere.

"Better turn us around," Ted LeFeuvre said.

"Hold on."

Torpey swung the cyclic. By the time he had steadied the aircraft and the Jayhawk's nose was pointed squarely into the wind, they were seven miles off the fly-to position.

"What's our airspeed?"

"Eighty-two knots."

"What's our ground speed?"

"Three knots."

"Jesus."

It felt as though they were riding a roller coaster, with rushes and sudden swoops and plunges, and each time the helicopter dropped sharply Ted LeFeuvre felt the sickening, hollowing-out, emptying sensation in the pit of his stomach that echoed on through his whole diaphragm. Torpey pushed the engines to 145 knots, and they began moving forward over the ocean.

"Hey, Steve," Ted LeFeuvre shouted. Everything was jangling around in the cockpit and he could not hear himself unless he raised his voice.

"What?"

"I was just thinking. You remember that really bad case a year ago that Adickes and Newby handled?"

"No."

"That real bad one. The one they did in the straits. I can't remember the name of it."

"The *Oceanic?*"

"That's it. Yeah."

"What about it?"

"Well, I remember Adickes telling me how bad they were getting beat up. Real bad turbulence. Just like now. Downdrafts. The whole works. And you know what they did?"

"No."

"They divided the controls."

Dividing the flight controls was not something pilots were supposed to talk about, let alone attempt, in the Coast Guard. It was an unorthodox technique, considered quite dangerous, and the command summarily discouraged the practice. The pilot was supposed to operate the cyclic, collective and floor pedals; the copilot was supposed to navigate, monitor the radio and program the H-60's flight computer. The rule book was clear on this. No dividing the controls.

But nobody's trained to fly in this, Ted LeFeuvre said to himself. These conditions are too extreme for one pilot to work all of the flight controls. Steve's already task-saturated. And we're just riding a tailwind. He's going to burn out fast once we start fighting the winds. But if one of us controls altitude and the other takes care of our lateral movements, we'll not only get crisper movement, we'll both last longer. Of course, we're going to have to make our control inputs jell. Can I keep up with you, Steve? I guess we'll find out.

"Want to try it?" Ted LeFeuvre asked.

Torpey hesitated.

"I'd take the collective and watch the radar altimeter. You'd stay on the cyclic and the pedals."

Torpey said nothing.

"It goes against the book," Ted LeFeuvre said. "It does that."

"All right," Torpey said. "Why not?"

"Good," Ted LeFeuvre said. "All right. Now, I'll follow your inputs on the collective with my right hand and gradually take it over from you. All right? You let me know what altitude you want me to keep."

"Roger."

"I'll do my best to maintain whatever altitude you want to hoist from."

"Sure."

"One more thing," Ted LeFeuvre told him. "When we get on scene let's just skip doing a PATCH." That stood for precision approach to a control hover. It was a standard approach that pilots used when they could not see the ocean below. "We'll probably get a visual of the water at a hundred and fifty feet, so why don't we just do a level-speed change. As soon as we spot the survivors, let's try to establish a hover anywhere between a hundred and a hundred thirty feet off the water."

"Let's go for it," Torpey said.

"Good."

Ted LeFeuvre thought of the air rushing at them as a kind of river. It is a very, very wide river, he thought, so wide that if we were in a canoe we would not be able to make out either shoreline from the middle. It's almost like we're salmon and this is the last river we'll ever be fighting against. It's a mighty river, this one, a surge of white water, and there are many big stones in the river. And it's a tricky river, he thought. It's fast and deep and traitorous and one you had better respect. It watches and waits until you think you have just about figured out its direction and then it swings on you, hard, when you aren't expecting it. And it always hits you from the angle you least expected. You respect this river, he told himself. Do not relax on it and take nothing in it for granted. That will be your end.

Steve Torpey instructed the crew to begin preparing the cabin for hoisting.

"Lee," Fred Kalt said, "start handing me those glow sticks."

"Here."

"And let's get the caps off a couple of flares."

"Okay."

While Kalt and Honnold cleared the deck and readied the flares and rescue basket, Ted LeFeuvre gradually took over control of the collective. At first he followed Steve Torpey's movements, touching the stick that controlled their altitude ever so lightly, petting it with his fingertips and the pads of his palm, and then gradually, almost imperceptibly, tightening his clasp on the collective until it was almost an extension of his arm and elbow and wrist and it was he who was moving them up and down, compensating for the plunging and upthrusting of the aircraft.

It *is* strange, he said to himself, this business of controlling half a helicopter. Now I know why they discourage us from doing it.

Most of his Coast Guard career, most of his life, in fact, it had always been his way to take total control of a situation, to shoulder all responsibility when confronting a crisis. And he had always relished playing the role of "Big Mom"—guiding, orchestrating, directing. This was different. Now he and a pilot nearly young enough to be his son were flying the Coast Guard's most sophisticated helicopter in a hurricane *together*. Each had the power to take the helicopter in a different direction, but they both had to come together to form a single pilot or they would go in the water. It had taken a freak of nature, an arctic hurricane of mammoth proportions, for him to realize that it was all right to relinquish control, that sometimes one man alone could not do it all. It was not an easy thing to admit; it ran counter to his training as a pilot, to his masculine ego. And yet it was liberating.

I've got to trust him, Ted LeFeuvre said to himself. And he's got to trust me. Considering how rusty I am in the H-60, that can't be easy for him. This is no absolute brotherhood. But we have to trust each other. We also have to read each other, time each other, anticipate each other. The seas aren't laying down. The wind certainly isn't letting up. I must say it feels

like we're flying in the fjords around Sitka, the way the winds are rico-
cheting around. It's like they're being deflected off the sides of mountains.
I guess they are in a way. Those waves down there are like small mountains. That's something to think about, all right. For now you better keep
that altitude steady around three hundred feet. That's enough to do for
now. That will be plenty.

He glanced back and saw Kalt and Honnold tying chemical lights to
the rescue basket. It made an amusing image—the silhouettes of two hel-
meted figures kneeling and hunched over a shiny metal cage, bathed in an
eerie, powdery green glow. It's like something out of a sci-fi movie, he
thought.

"Fred," Ted LeFeuvre said over the ICS, "it looks like you guys are
dressing up a Christmas tree."

Kalt grunted. "We're going to make *sure* those survivors see this stinking
basket."

"How could they miss it?"

"It'll be their fault if they don't see it. Lee, give me another glow stick."

Right then the direction finder began swinging.

"I think we got two signals," Ted LeFeuvre said to Torpey. He squinted
at the dial. "Yup. We got two."

"Two?"

Sometimes the needle would swing hard left, and other times hard
right. Sometimes it squirreled from side to side. We must be picking up the
EPIRB and the data marker buoy left by the first aircrew, Ted LeFeuvre
was thinking. It looks as though they are three miles or so apart. One of the
beacons must have drifted faster than the other. Well, which one do you go
to first? Are the survivors closer to the EPIRB or did they get swept along
with the data marker buoy?

Which signal is which?

"If you ask me," Ted LeFeuvre said to Torpey, "I'd say the DMB is the
one off to our right."

"Why?"

"I don't know. Just a feeling."

"So which one do we head to first, sir?"

Ted LeFeuvre studied the dial again. It was a fifty-fifty guess, and he did
not like guesses.

"Let's take the one to the left," he said. "I'd say it's about three miles away."

"All right," Torpey said. Then, over the ICS: "How are you guys doing with the flares?"

"Not so good," Kalt said. As soon as they had turned into the wind the ride had gotten a lot rougher and equipment was sliding and tumbling and sometimes flying back and forth across the cabin. They had not yet armed the flares; in fact, they had not yet removed them from their Styrofoam casings and nylon bindings.

"Well," Torpey said, "hurry it up."

"We're working on it."

To Ted LeFeuvre, Torpey said: "I'm going to descend to a hundred and fifty feet."

"Roger that."

Until then, they had only snatched glimpses of the waves. But now, as they descended, Ted LeFeuvre had the feeling of seeing something that had begun normally grow quickly into large, then oversize, then gigantic proportions. It was as if they had tossed a rock into a pond and the ripples had multiplied and magnified in size and had rebounded right back at them as a tidal wave. It had been almost impossible to see through the sleet and snow that overtasked the wipers. But as they dropped down closer he could see the ocean heaving, splitting and pulling apart in craters.

So that's why the beacon signal keeps coming in and out, he said to himself. The waves are blocking the signal. They block it each time the EPIRB skids down into a trough or gets swamped by a wave. Those seas must be huge. Well, I'm not losing the beacon signal anymore. I've got a GPS fix on it.

I hope those guys hung on to the EPIRB, he thought. If they let it go we won't ever find them.

The helicopter was bouncing off gusts but crabbing forward ever so slowly. Ted LeFeuvre was squinting and scanning the blackness, hoping for a glint or a flash or anything that would give them something to hone in on. I hate this part of a mission, he thought. It's the uncertainty. Where *are* they? Are we going to find them? Are they dead already? What can we do better? Change our search pattern?

Behind him, Kalt was talking to Honnold.

"Listen," Kalt said. "The captain and Mr. Torpey are really busy up there. They might not see the survivors."

"I know."

"We gotta open the door *now*."

Normally the flight mechanic asked the pilots for permission to open the jump door. But Kalt just strapped on his gunner's belt, double-checked that it was attached to a cabin hook, slid on his knees over to the door and threw it open without a word.

A blast of flying ice hit him with the force of a fire hose. He leaned back on his heels, mopped his visor.

"Lee," he shouted. "Get that Maxi-Beam and point it out here, will you?"

Honnold grabbed the handheld light and aimed it down at the ocean.

"That's it. Over this way."

When the door opened, Ted LeFeuvre felt a lick of cold air across the back of his neck, and swiveling, seeing the jump door up, Kalt and Honnold swinging outside in their gunner's belts, slashing at the swirling, flying sleet and snow with the light of the Maxi-Beam, he thought: Good move, guys. Good move. We should have thought of doing that earlier.

The torch lighted up the ice, but sometimes punched a beam through it all the way down to the ocean surface. In the light, the sea looked like it was smoking, rolling over, boiling. At times they could make out a wave below and aft of the Jayhawk, and sometimes they could see a wave before the nose of the helicopter. But sometimes they saw nothing at all. There was no pattern to it.

For several minutes Kalt crouched on the lip of the jump door, the sleet rattling on his visor and helmet, the roar of the wind and turbines in his helmet.

Then he looked up at Honnold and said, with a flat, emotionless voice: "I see them."

Fred Kalt had that Steve McQueen look: chiseled jaw, sharp nose, cool, steady eyes as blue as ice, not a trace of anxiety in them.

He was sturdy but not burly, and not particularly tall. He kept his hair, hickory brown and touched with gray, cropped and standing at attention. When he smiled, it was a bemused, boyish smile. When he spoke, his words rolled out calmly, gently. When he laughed, others laughed with him.

It was a rare day when Kalt got angry on the job. Club Fred. That's what they called him on the hangar deck. The man least likely to spill his popcorn at a horror flick.

Born on Long Island, raised in Florida by a woman who divorced when he was eight, Kalt entered the air force at eighteen. He wed his teenage sweetheart, moved with her to Syracuse, New York, left her—he walked in on her at home while she was cheating with his buddy—allowed his four-year enlistment to run out, returned to Florida and went to work delivering packages for UPS. A year later he fell ill, a virus, and as he had gotten in the habit of seeing the base doctor for even the most minor of ailments, he skipped the home remedies and saw a physician. A week later the bill came in the mail: four hundred dollars. For the first time in his life, civilian Kalt had to pay out of pocket for a doctor's visit. It got him to thinking. Within six weeks he was a Coastie.

In 1989, Kalt was the honor graduate at the guard's Aviation Electronics School at Elizabeth City, North Carolina. He was also a husband again—this time to a woman named Barbara. She had a daughter, Jessica, they had another, Kelly, and then Kalt got shipped out. During the next seven years the Kalts bounced around, from Oregon to Florida to Massachusetts. In 1996, he put in for a transfer to Alaska.

On their second night in Sitka the Kalts were lying in bed, listening to the local radio station, when they heard a report that authorities had just closed the Totem Park trail, just outside of town. Black bears were afoot. Kalt quietly clicked off the radio. His wife shut her eyes, shook her head and said how she wished she were back in civilization. It was June 1996.

Gee whiz, he thought. What do I do now? Get down on my knees and beg her to like it?

Now, eighteen months later, eight days shy of his thirty-fifth birthday, Kalt *was* on his knees—on his knees in the back of an H-60 helicopter that was getting mauled by an Alaska hurricane, and not liking it too much.

"Fred," Steve Torpey was telling him, "we'll start by making two flare drops. Let's dump the first bunch about a hundred and fifty or two hundred yards upwind of the survivors. Then I'm going to let the helicopter slide backward a little bit. Then I'm going to take us forward and right, and then we're going to toss some more smokes out at the two o'clock position."

"Roger."

"Just remember," Torpey said, "everything you got you toss as soon as we get to the position."

"Roger."

Kalt stuck his head outside.

The strobe had slid beneath the helicopter. Around it, glinting in the beam of the searchlight, he saw a gaggle of reflective tape. There could be two survivors, he thought. There could five. Whatever. They're alive.

He tried shucking the Styrofoam off the flares, but he could not free the Mark-25s from their casings. Lousy manufacturers, he thought. Just like a ketchup bottle. You've got to be a genius to open one.

Then it came to him.

He pulled out the cutters he normally used in emergencies to sever the hoist cable, snipped the nylon bands and was tearing the Styrofoam casings off and arming the flares when he heard Torpey say:

"Stand by to deploy flares."

"Flares are ready."

"Drop! Drop! Drop!"

Just like a lineman sends a football through his legs to the holder of a field-goal kicker, Kalt snapped one, two, three flares between his legs and out the door.

"Flares away!"

He spun around and leaned outside. Down below, the flares shot red-white flames across the black water.

"Flares are in the water. Flares have ignited."

In the cockpit, Steve Torpey saw none of it. Sleet was blanketing his windscreen and everything—the horizon, the sky, the water—had whited out. Oh shit, he thought. This is bad. This is real bad. He looked blankly at his instruments. They seemed foreign to him. He had been flying on visual cues for a while now. There was a mental transition he had to make so that he could start assimilating and interpreting the information on his panel. I don't like this. His gut was coiling up. Calm down. He eased back on the cyclic just a hair, to let the helicopter slide back a little.

When the gust caught the rotor system they were instantly thirty degrees nose up and backing down. Ted LeFeuvre had no time to read their rate of descent; he had only enough time to react, to pull on the collective.

The radar altimeter was unwinding fast.

We're backing down.

Torpey had the cyclic all the way to the dash to level the nose.

We're still backing down.

Hearing only the high-pitched, whining growl of the engines, feeling the awful hollowing, scooping sensation in his stomach, the floor of the helicopter seeming to drop out from under him as it went down, down, faster and faster in a backward, plunging rush, his helmet pinned back against the headrest so tightly it felt glued to it, he knew the pace was too much. But he held the collective. He would not let go of it. They were flying backward and he had no idea what was behind them and still he held the collective back, feeling the backward-rushing sensation, as though he was in a station wagon without a rearview mirror, racing in reverse down a hill at 100 mph.

Then came the screams.

"UP!"

"ALTITUDE!"

"EMERGENCY UP!"

And then he saw the wave through his windscreen.

It was all black, except for the white line of drool along the top, and was closing and building with a petrifying smoothness of motion. When it was within fifty yards and Ted LeFeuvre saw the steel, cylindrical casings of the flares embedded in the wave, spinning, not going end over end, but spinning and shining silvery in the bright white light, he squeezed the collective stick harder, his hand now cold-drying wet, his eyes locked on the smoothly coming darkness.

"UP!"

"UP! UP!"

"UP!"

Ted LeFeuvre was trying to be loose but steady, trying to hold his breath and not think about anything but the radar altimeter; to focus and keep his cold-clammy hand on the collective and maintain distance from the wave. The radar altimeter read forty feet. Seconds passed.

The altimeter still read forty.

This can't be, he said to himself. I'm pulling this helicopter up at *full power*. We should be going *straight up*.

Then it hit him.

They *were* going straight up. But below them, the wave was rising at the same speed.

Well, Lord, Ted LeFeuvre thought, I am going to meet You now. But, Lord, do I have to go out being cold and wet? You know how I hate cold and wet.

At that instant the helicopter lurched skyward. The rogue wave broke just beneath them.

By the time Ted LeFeuvre was able to arrest their ascent, the Jayhawk had climbed to six hundred feet above the ocean and sailed a mile downwind of the survivors. It took them another ten minutes to get back on scene.

The Mark-25s were still visible, upwind of the strobe light.

"Okay, guys," Torpey said over the intercom. "Get those smokes ready. And this time, Fred, don't use any of those small flares. They burn out too fast. From now on, all that goes in the water are the big ones—the Mark-58s. Got that?"

"Roger."

Torpey looked at the torque gauge.

It measured the stress on the engine's gearbox—stress caused by pulling too much power, too suddenly on the collective or cyclic. The H-60's maximum acceptable torque was 127 percent; anything above that could cause *overtorque* and rupture the transmission or lubricant seals.

They had pulled 132 percent torque.

"Captain," Torpey said to Ted LeFeuvre. "Did you see that? Did you see the torque gauge, sir?"

"I saw it, Steve."

"What do we do, sir?"

"Keep flying." Ted LeFeuvre made his voice sound disinterested. "Just keep flying."

Torpey went back to work. His movements were as crisp as they had been at takeoff, Ted LeFeuvre thought, as crisp as a machine, and he had no trouble banking the helicopter in a two o'clock position above the survivors. They dumped another seven Mark-58s.

"That was good," Torpey said. "Okay, let's complete part two of the rescue checklist. We're going to do a basket hoist."

He moved the Jayhawk to his right and slowed his airspeed. The wind shoved the helicopter back. Now they could approach the survivors from the front.

Lee Honnold unhooked the rescue basket from the cargo straps and set it on deck. Fred Kalt slid over to the winch. Ted LeFeuvre flipped up two toggle switches on the console above his head, supplying power to the hoist.

"Backup pump is on," he said.

"Roger that," Kalt said.

Torpey said, "Okay, listen up."

He told the crew to be ready to hoist from an *average* altitude of 110 feet and to dump flares on each approach to the water. Then he reminded Kalt to fully secure the bags of lead shot in the rescue basket so they didn't get slung back into the tail rotor.

"That's it," he said. "Any questions?"

"Mr. Torpey?" Mike Fish, the rescue swimmer, broke in. "You want me to get dressed out? I'm willing to go, sir, if you want. I feel I can do this."

"Hold on, Mike."

Torpey muted his ICS line and turned to Ted LeFeuvre. "You know, Captain," he said, "I don't think there's any way we should put a rescue swimmer down in that."

"No," Ted LeFeuvre said, softly. "I think we should start with just the basket, too."

"Okay."

Torpey clicked on his ICS connection and said, "Sorry, Mike. I don't think we're going to use you."

I wonder whether the other rescue swimmers offered to go in, Ted LeFeuvre was thinking. I'll bet they did. To help somebody they never met. These swimmers are young guys. With families. Mike here's got a wife. If he doesn't come back, she's nobody's wife anymore. Nobody thinks of that. That is selflessness. On her part as much as his.

Fish said, "I just wanted to let you know that I'm ready to go if you need me to go."

"Mike," Ted LeFeuvre said, "for now, just keeping working comms.

Back me up on the radar altimeter, too. Call out altitudes if we go below eighty feet."

"Roger."

"Fred?" Torpey said to Kalt.

"Sir?"

"Get ready to work with me now," Torpey told him, "because you're going to see some pretty big changes in the way I'm going to fly this thing."

F O R T Y - S I X

The rescue basket was penduluming now beneath the helicopter. Each time it swung the hoist cable dug deeper into the rail below the door frame and chafed on the Night Sun.

Fred Kalt just watched it swing and swing and swing until, finally, a wave smacked the basket into a trough and buried it under a cascade of water.

"Is it in?"

"It's in."

"Basket's in the water!"

Kneeling, the sweat running down his back, the flying ice sniping at his cheeks, Kalt watched the green glow of the chemical sticks fade as the basket settled under the waves.

"Hey," Kalt said. "I think I got it close to them this time."

He cleared his visor of sleet and looked down. The basket had resurfaced again. The green glow was only about five yards from the flashing strobe.

"Why aren't they climbing into it?" Lee Honnold said. He was lying spread-eagled on the deck, shining the handheld searchlight on the survivors.

Kalt kept his eyes on the glowing basket.

"Shit," Honnold said. He was breathing heavily. "It's right there. It's right there in front of them."

"It's sinking below the surface," Kalt told him. "They can't see it."

"They can't?"

"No," Kalt said. He was thinking that he had never really seen waves before. "The weights we put in the basket are dragging it down under the water. I'm going to pull the basket up."

"Shit."

Kalt threw the winch in reverse. They had been hoisting more than forty minutes. The first few drops had been almost laughable. But the next ten tries Kalt had improved his aim, so that now he was dragging the basket to within five yards of the survivors.

"Here it is," he said.

The basket was pitching just outside of the jump door. He reached out and pulled it in the cabin. We might not get them, Kalt was thinking. Hell with that. We'll get them. We've just got to make an adjustment.

"Okay, let's get rid of these weight bags," he said. "Give me a hand unclipping them."

While they got the basket ready again Steve Torpey was working as hard as he ever had in any helicopter. The H-60 could be programmed to fly all itself, but the computer was all but useless now. Torpey had to whip the cyclic around and pump the tail-rotor pedals like a madman just to keep them reasonably level. He was doing thirty-degree-angle banks, lifting the helicopter's nose up, throwing it down, wrenching it hard right, left, then left again, then hard down, up, right, left, back, and doing all of this while changing his thrust vector to compensate for the gusts. He was also suffering from what pilots called a helmet fire. The inside of Torpey's helmet was crawling with so much sweat he had even turned on the cockpit's air-conditioning.

Relax, he was saying to himself. Just take it easy. You're squeezing the black out of the cyclic. The harder you grip the cyclic, the worse you fly. Relax your grip. And breathe. Yes, breathe. Just breathe.

"Hey," he heard Ted LeFeuvre say, "those Mark-25s are starting to go out."

"We better get some new ones out there." Over the intercom, Torpey said, "Hey, Fred. Let's get out another eight more of those Mark-58s."

"Roger that," Kalt said.

Mike Fish held the flares and Honnold cut the bindings. They pulled the cylinders out of the canisters and handed them to Kalt, who turned their tabs to arm them and then lay them out on the deck perpendicular to the door.

The helicopter took an uppercut from a gust and the flares hopped around the cabin.

Kalt made a grunting noise and, shuffling with his hands on the deck, picked up the flares and then cradled them to his chest.

"Okay," Torpey said, "prepare deploy flares."

"Hold on," Kalt said.

"What's wrong?"

"Nothing," Kalt said. He could feel the helicopter shooting another approach to the water now. "Almost ready."

"All right," Torpey said. Then, as he came up on his twelve o'clock: "Prepare to deploy flares. Okay . . . Drop! Drop! Drop!"

Kalt counted to two before pitching the flares out the door. He wanted the flares spaced farther apart in the water to increase the length of their visual reference. The flight mechanic stuck his head out the cabin and saw the flares burning in an arc just thirty yards upwind of the strobe light.

"Flares have ignited," he said in the same deadpan tone of voice that had intensified only slightly since they had departed Sitka. "They're just upwind of the survivors."

"Good."

As they were dropping the flares, Ted LeFeuvre had been trying to hold their altitude steady and make some sense out of the radar altimeter, which was swinging between 70 feet and 220 feet. Initially, he had backed up Torpey on the cyclic. But the inputs were changing so quickly and so extremely that he feared that just having his hand on the cyclic might slow Torpey down. So Ted LeFeuvre kept his hand on his thigh and watched as the young pilot steered, though he reminded himself to stay awake and be ready to take over the moment things went awry.

Something deep inside was urging him to seize control of the aircraft and to fly them far away from there. But he fought the urge off.

We're doing all right, he thought. We're doing better than I'd expected. The kid is flying brilliantly. There aren't many people who could fly like this kid is now. I wish I could. I wish there was some other way I could help him. Well, you're helping him plenty. Just let him go. And keep us out of the water. Do that.

"Captain LeFeuvre," he heard Mike Fish say. "Watch our altitude, sir. We're now at seventy-two feet."

"Thanks, Mike." Ted LeFeuvre pulled gently up on the collective. "Taking us back up."

This is just like being inside the motion simulator at flight training school in Mobile, he thought. The way you see nothing but black, then a few flares rising up the walls of the waves, then nothing again. But here, he thought, you don't get a chance to take a break. There are no breaks. No lapses, either. No, you aren't allowed any of those.

Sometimes he could see tremendous streaks of foam being ripped off the wave crests and slung in long, white lassos, and he noticed that Torpey was using those foamy streaks as references and angling the helicopter so as to keep the wind planing off the aircraft's nose. Then everything would go blindingly white again, and he would have only the radar altimeter to focus on again.

But now, for the first time since they had left Sitka nearly two hours earlier, he was having trouble keeping his thoughts from wandering. He scolded himself, but he found that no matter how hard he tried he could not stay focused on the altimeter and its grave importance. I wonder if I left enough pellets in the stove tonight, he thought. Gosh, it'll be a real pain if I get home and the house is freezing again. I wonder what Cam is doing right now. Probably sleeping. He better be. Kathy lets that boy get away with murder. She would never listen to me about that boy, though. Gee, that was a big wave. Hey. What is the matter with you? Watch your altimeter. What does it say? Eighty-nine feet. Pull up. Get us back over a hundred. That's it. I'm too old for this. I'm way too old for this. I should be back in Sitka writing the award citations for these guys. I shouldn't be here. That's it in a nutshell. I should not be in this seat. I wonder how I would have handled this when I first came up as a pilot? Stuff like this is for the young guys. This is going to be a heck of an award write-up.

Good grief. What are you thinking? Oh, no. Watch that wave. Pull us up. Ooh. That was close. That was close that time.

Over the intercom he could hear Kalt mutter, "Oh, oh." The flight mechanic had just pitched the rescue basket out the jump door again.

"Mr. Torpey?" Kalt said.

"What?"

"The basket is sailing from side to side." Kalt was hanging halfway out the door. They could hear his mouthpiece picking up the wind's howl.

"The basket is flapping in all of this wind. It's sailing aft at forty-five degrees."

"What can I do?"

Kalt pulled himself into the cabin. "Let me get it in and try again." He threw the winch in reverse. "Go forward and right," he said. "Forward and right—*two hundred.*"

They tried turning the helicopter a bit to create a lee but that did not work either. Over and over Kalt threw the basket out, hoping the gusts would stop, trying to time it so that the wind would not fling the cage into the tail rotor. A few times he did get the basket to drop beneath the reach of the rotor blades. But those times the basket did hit water it bounced and twirled from crest to trough, appearing and disappearing in the foam-laced swells.

After a half hour of attempts Kalt said, "Mr. Torpey, I can't seem to do this hoisting, sir. Do you want to see if maybe Lee could do it better?"

All this time Honnold had his six-foot-two-inch frame sprawled stomach down on the deck, his long arms and legs clinging to the floor and the door frame as though he were riding the back of an enormous, bucking bronco. While Kalt worked the winch, Honnold had been sparring with the hoist cable to keep it from rubbing on the helicopter and snapping. His head was ringing, the deck beneath him was pitching and rolling, and he felt sick at his stomach. His arm muscles burned and his hands were cold and losing feeling. He stuck his head out the door, into the torrent of lashing ice, seized the cable and pulled it up on his helmet so that his head served as a guiding shaft; with the line running off his helmet, it no longer caught on the door frame or the searchlight. But each time the helicopter jerked, the great weight of the steel line gouged his helmet and wrenched his neck.

He was grunting and swearing now.

Torpey heard it. "No, Fred," he said. "We're going to stick with you for now, okay?"

"Roger that," Kalt said, and he leaned back out the door. "All right, let's go forward and right a hundred and fifty."

Torpey was really laying into the controls now, no longer banking twenty

or twenty-five degrees, but now routinely banking the helicopter at a forty-
to forty-five-degree angle. At one point, Ted LeFeuvre began to wonder if
the tips on the rotor blades might not snap off under the strain.

But it began to make a difference.

Torpey and Kalt found a rhythm and soon the conning commands
were not as dramatic: Kalt spoke almost softly, like a surgeon, talking his
pilot through the evolution as calmly as if they were setting down on a
deserted beach, telling him to go fifty feet this way, thirty feet that way,
twenty feet aft, fifteen forward, until they were consistently within a tightening area. The helicopter was still heaving, pitching wildly in the wind;
but it was no longer sailing all over the sky.

Kalt was saying, "That's it. That's good. Now back and right ten feet."

"Back and right, ten," Torpey repeated.

"Forward five."

"Going forward five."

Down in the churning sea, the basket was bobbing within ten yards of
the survivors now.

"I've got the basket near the survivors," Kalt said, in that same, flat tone.
"Paying out slack . . . okay . . . Lee . . . give me some light . . . right there
. . . hold it . . . that's it . . . *hold!*"

Torpey laughed.

"Hold? In this?"

"*Hold!*"

Kalt could only see blurry shapes in a circle. Then one of the shapes
broke from the others and made for the basket. He saw the flash of reflective
tape.

"Someone's swimming toward the basket!"

Grabbing the hoist cable now, feeling the heavy tautness of the steel
fibers sliding through the fingers and palm of his leather hoisting glove, he
waited for a tug in the line.

Then:

"I think I got somebody! Yes! We got one in the basket! Taking a load!"

Roughly a hundred feet down from where Kalt was kneeling, Bob Doyle
was shouting to Mark Morley:

"Mark, I'm cutting you free of the rope now!"

"Just get me close! Just get me close! I'll get in the thing! I swear it!"

"Okay, Mark. Take it easy. I'll get you there. You're the first one up, okay?"

"Where? Where?"

"It's close by. Close. See what I told you? You're going to see your kid."

Once Bob Doyle had heard the distant throbbing turn to a whining roar and seen the spotlight, he felt the hopeful, singing feeling around his heart, and then the helicopter was overhead, much lower than the others, and then shoots of bright, white light were bursting around them and casting shadows and lighting the waves green again, and then he saw the glint of the hoist cable in the coned light of the belly floods. God, don't let the waves touch them, he thought. *Please* don't let it happen.

Then the helicopter went hurtling downwind.

He had watched it go shooting away until it was almost out of sight. Then he had seen it coming back. It had wobbled up from horizon, growing bigger and brighter, and then he saw the shine of flare casings tumbling through the sky and more bursts of the red-white light not far off, and then waves avalanching over the glowing basket, but the basket moving closer, all the time closer, and then he was thinking: God, bring it to me. I'll grab it and I won't let it go. I swear. And then, hearing a wave gathering in that same sickening, gurgling rush high above, seeing nothing and breathing nothing and knowing nothing, tumbling in hard, heavy darkness and then snapping free of it, mopping his eyes, he spotted the glowing green rescue basket no farther away than the length of two swimming pools.

Bob Doyle yanked his suit zipper down to his waist and, feeling the icy shock on his chest, pulled out his fishing knife.

"Mark," he said, "I'll get you in the basket. Two people can fit in that basket. When we get to it, you grab it. You hang on. Even if I can't get in."

"I gotta get in it."

He cut the rope around Morley's waist. "I want you to swim as hard as you can." He severed his own line. "I'll be holding you." He let the knife go. "Giggy, I'm taking Mark up!"

"Go!"

Reaching his arm over the skipper's back, Bob Doyle started kicking and thrashing. Every muscle felt rigid. Needles of pain shot through them. He swam hard. It did not seem like he was moving. The waves kept pushing him back. He had Morley by the collar. The skipper was trying to do a dog paddle and not doing it too well. They took a wave in the face. Bob Doyle spit it out.

Ahead the green glow was rising and falling in the blackness.

"Move!"

His legs felt like lead. The glowing box was coming straight at him. He swam as hard as he could. The swells were lifting him up and down, but the glow was brighter and brighter. He felt a sharp pain on his skull.

Bob Doyle grabbed the metal cage with his free, left hand, steadied it.

"Mark! Get in!"

A swell lifted them high, high, high, and then dropped away and they rolled down its back side. Bob Doyle still had the basket in his fingers. With his free arm he reached around Morley's waist.

"Get in!"

He tried to heave him into the basket. He got behind him and tried to push him by the rump.

No good.

"Christ," he screamed at Morley. "Help me!"

Get leverage, he thought. Get in the basket and pull him up and into it.

"Here," he shouted into Morley's face. He grabbed the loose, heavy arms, draped them over the top of the wired basket. "That's it. Now hold on to the cable."

Bob Doyle swung around in the water, grabbed the opposite side of the basket and hoisted himself up and in so that his knees pressed off the bottom of the basket.

"Come on!"

On his knees, his hands grabbing Morley's, he pulled with everything he had.

"Come on!"

Again he struck back hard against the great weight.

"Get in here!"

Just then he felt a heavy jerk.

———

As soon as Ted LeFeuvre heard Fred Kalt shout that there was a man in the basket, he pulled full power on the collective. The helicopter shot skyward.

That's it, Ted LeFeuvre was thinking. No faster way to lift that basket out of those waves. The hoist can't lift that basket like I can. Oh, boy, are we going up.

Lee Honnold had been thrown aside when the helicopter lurched skyward. The cable had whipped down on the deck. It was now sawing against the door frame and along the grating of the Night Sun, the searchlight on the belly of the helicopter.

Fred Kalt had been catapulted backward. He peeled himself off the back wall and staggered to the door.

Below, the basket crashed through a comber and, spinning and shedding foam, punched through the far side of the wave.

"Holy crap!" Kalt shouted. "The survivor's still in the basket!"

The winch was taking cable onto the reel in sweeps as fast as the reel could turn. There was no slack at all; the cable was as taut as a tuned fiddle string.

"Basket's halfway up!"

The cage, tiny at first but growing steadily in size, pitched and spun, engulfed in curling curtains of sleet and snow.

"Basket's twenty feet below the cabin!"

Up, up, up it came, until it swayed just outside the jump door.

"Basket's outside the cabin door!"

Kalt reached for it. The basket swung away from him.

"Bringing the basket in!"

This time he grabbed the metal cage and pulled. It didn't budge. He pulled again.

Stuck.

"Bringing the basket in!"

He yanked harder.

"Attempting . . . to . . . bring . . . the basket . . . in," he said, grunting. "It, ah . . . it . . . the basket won't come in the door."

Steve Torpey glanced at his rearview mirror. Outside, through a heavy shroud of driving white flecks, he saw the rescue basket. There was a

crouched figure inside it. He also saw a flight helmet. And he saw two arms
pulling at the cage.

Kalt was crouching now at the door, shouting to Honnold, *"Pull, Lee!
Pull!"*

Both men were now leaning back, pulling with all the strength in their
cramping muscles.

"Are you pulling?"

"I'm pulling! I'm pulling!"

"It's not coming in!"

"I'm pulling as hard as I can!"

Mike Fish, in his seat monitoring altitude and working the high-frequency
radio, looked up. Through an opening between Kalt's right leg and the
jump door, he saw why the basket would not enter.

A second man was dangling from it.

Each time Kalt and Honnold tried to yank the basket in, the dangling
man's arms and head got rammed against the lip of the jump door.

F   O   R   T   Y   -   S   E   V   E   N

Through the driving sleet icing his visor, Fred Kalt could hardly make
out the kneeling figure inside the rescue basket just outside the helicopter
door. He reached out and gave the basket another yank.

Something was wrong. No matter how hard he tried, he could not pull
the basket inside.

"FRED!" Mike Fish shouted, behind him. "SOMEONE'S HANG-
ING ON THE BASKET!"

"Where?" Lee Honnold screamed into the wind. "Where? Where?"

"FRED!"

"I can't see him!" Kalt shouted.

The man was inches below Kalt's boots, barely clinging to the bottom
of the basket. He lifted his head up, looked into the cabin and locked eyes
with Fish.

For a second. Just one second.

Time enough for everything to pause in Fish's mind, for the whining sleet and the groaning turbines to hush.

Time enough for one man's eyes to scream for mercy, for another's to scream in horror.

And then he was gone.

Not a minute earlier, the basket was eighty feet below the helicopter, bouncing like a yo-yo in the wind and the whirling, thick snow and sleet.

"We're getting there!" the man on his knees inside the rescue basket was screaming. "Just hang on!"

The man dangling from the bottom of the basket yelled back: "Hang on to me!"

"I got you!"

"Don't let me go!"

"I said I got you!"

Then there was a hollowing swish through a rushing noise, and then a deafening, rumbling roar, and a wave exploded in a cloudlike burst only feet below the dangling man's legs. The blast from it rolled up and buffeted them.

"Don't drop me!"

"I got you!"

The man kneeling inside the basket, Bob Doyle, had his hands under the armpits of the dangling man, Mark Morley, and he was saying to himself: We're going to be okay now. The sea can't get us anymore. *We're out of it. We're out of it.*

The basket kept spinning, twirling, shedding spray and spindrift.

"We're almost there!"

"I can't hang on anymore!"

"Give it what you can!"

"I can't!"

"Don't let go!"

"Please don't drop me! Please don't drop me!"

"You ain't gonna drop!"

"Don't drop me!"

"I won't!"

They were in the belly lights of the helicopter now. They were fifteen feet below the jump door, and as they climbed Bob Doyle saw, out of the corner of his eye, helmets and shoulders hanging out the side of the helicopter.

"Don't drop me!"

Just then a gust slammed into them. The basket rocked and whirled. Bob Doyle's hands no longer had his skipper by the armpits; they had slid down his arms and were fastened to Morley's wrists.

"Hang on!"

Morley's hands, which had been clutching the basket, were sliding now.

"Don't let go!"

Bob Doyle lunged with one hand and grabbed his skipper's collar. Leaning back, knees digging into the wire mesh of the basket bottom, he swung his other hand around and seized the shoulder. Now he leaned back.

"Bob!"

"I got you!"

The upper half of the basket was now above the deck of the helicopter cabin.

"We're here!" Bob Doyle screamed hoarsely at the shapes in the doorway. He looked down.

"Hang on, Mark! We're here!"

"I can't!"

He thought he could see Morley's eyes now; they were blank eyes, black as coal, all fear and despair and hope drained and emptied out of them from the fight. There was nothing in them at all. They were bottomless.

The basket lurched.

"Hey!"

There were now two pairs of gloved hands yanking at the basket frame. He tried to shout but the groaning roar of the turbines and the whining sleet swallowed his screams.

"*No! Wait!*"

Another lurch; this time he saw it. The head of the dangling skipper rammed against a steel rail beneath the door frame.

"*No!*"

The basket was wobbling.

"*No!*"

Again the basket lurched. Again Bob Doyle heard the dull, sickening thud of Morley's head against the fiberglass airframe. This time, Morley lifted his head.

He turned it a little to the left, then turned back and looked straight up and locked wild eyes with the bearded, screaming man in the basket above him.

His friend.

"*No!*" the man, Bob Doyle, was shrieking. "*Oh please, Mark . . . don't . . .*"

And then Mark Morley allowed the wind to take him in any direction that it wished.

The altimeter on the central display unit read 103.

One hundred and three feet, Mike Fish was thinking. My God, that's far for a man to fall.

The clunk of steel on the deck snapped Fish out of his thoughts. It was the rescue basket. Fred Kalt and Lee Honnold had finally pulled it in.

"Basket is in the door!" Kalt shouted. He was elated; they finally had a survivor in the helicopter. There was nothing that could go wrong now.

The man in the basket was hysterical, gesturing, blubbering. Honnold was trying to calm him down.

"What the hell's wrong with this guy?" Honnold said. "He's going frickin' nuts."

"Fred," Fish said. He felt a sinking feeling going all through him. "Fred?"

Kalt didn't hear him.

In all of the confusion, the ICS cord plugged into his helmet had come loose. He could not hear his crewmates. Fish got up and tapped him on the back and Kalt turned around. Fish pointed to his own helmet.

Kalt understood. He picked the cord up from the deck and plugged it back in to his helmet.

"What's the matter?"

"There was someone hanging on the basket."

"Are you sure?"

"He just fell."

Kalt whirled around. "Where?"

"There. He was just there. He fell."

Kalt just stood looking out the jump door, looking down at the seas.

Beside him, Honnold tilted the basket and the survivor rolled out. The man crawled to the back wall and leaned his head against it. Honnold pulled the man's hood off.

"You all right?" he shouted over the drone of the rotors.

"The skipper," the man shrieked. Tears streaked his reddened cheeks. "The skipper just fell. Oh, God, I let him go! I let him go!"

"Who?"

"It's my fault! I let him fall! I couldn't hold him!"

The man broke down, sobbing.

"Hey," Honnold said, confused. He grabbed the man by the shoulders and head-butted him.

"Calm down!"

He head-butted the man again, and shouted: "Listen to me!"

The survivor stared almost fearfully at Honnold. He did not want another head butt.

"You are one lucky bastard!" Honnold yelled at him. "You know that? Do you? Now calm down. Calm down and tell me how many people were on board your boat."

The man wiped his eyes. "Five."

"Not four?"

"Five."

"You say somebody fell? Who fell?"

"Mark Morley, our skipper."

"Is there anyone else down there who's alive?"

"Yeah," the man said. His lips were trembling. Icicles twitched in his beard.

"How many are down there?"

"Two," the man said. His voice was hoarse, heavy. "Plus the skipper."

"Okay," Honnold said, helping the man up. "Here, get in this seat. Easy now. I'm gonna strap you in."

Tears ran down the man's cheeks.

"Jesus," Honnold said, "that's some mess you got yourself into. It's a real mess down there."

The man wiped his eyes and looked at the smoke flares on the cabin deck. "You," he stammered, "you—you guys need any help with that stuff?"

"No," Honnold said. He was pulling a strap over the man and buckling him in. "Just take it easy."

Mike Fish crawled over with a thermal-insulated sack. He was responsible now for checking the survivor for hypothermia and shock. He began to scrutinize the man's flushed, drawn face, and then froze.

"Hey," he stammered. "It's . . . it's . . . *Bob Doyle!*"

Over the intercom, he heard Steve Torpey break in, "What? *Our* Bob Doyle?"

Fish reached over and touched Bob Doyle's beard. "Yeah," Fish said, and then to Doyle: "Bob, it is you, right?"

"Yeah, it's me."

"What the hell's he doing out here in this?"

"Wait," Fish said. "He's . . . He's saying something about the skipper falling . . . He says the boat's skipper just fell . . . That's the guy who just fell off the basket. Sir, he's pretty upset."

"Well," Torpey said, "just tell him we're going to do our best. Tell him we'll get to him."

"Roger."

In the cockpit, Ted LeFeuvre was working the collective and watching their altitude on the radar altimeter. He could not keep from hearing their talk. But he had not taken his eyes off the console or the seas, not even when he heard the commotion over the fallen survivor. He wondered how it must be to fall through darkness and not to know when you would hit the water but that distracted him momentarily so he told himself to stop thinking about that. He was having a difficult time as it was concentrating on his inputs and keeping the aircraft in a hover. But when he heard Fish say over the ICS that the man they had rescued was Bob Doyle, he turned his head with a sort-of lunge, stared at the shadowy figure slumped against the back wall and turned back.

I can't believe it, he thought. We saved Bob Doyle? The guy who caused me so much grief? Throw him back. No, no, no. That's horrible.

That's bad. Lord, I'm sorry. Truly, I am. But, Lord, of all people—Bob Doyle?

Fish was checking his vital signs. "Mr. Doyle," he said, "what are you doing out here?"

Bob Doyle thought it felt good to hear someone call him mister. "I'm just glad to be here," he said, and he sighed. Then he smiled and shook Fish's hand. "Thank you, Mike. Thank you."

"That's all right," Fish said. He kept working on him. "Are you cold?"

"Very."

"Do you want me to put you in the thermal bag?"

"No, no," Bob Doyle told him. "I don't need that. Save it for the other guys."

"Are you sure?"

"Yeah."

"All right," Fish said. "You just relax, sit back and enjoy the flight."

"Just get the skipper."

"We will."

The basket was already going down again. It splashed in a trough between two enormous waves, ten yards from the survivors. Kalt was watching the cable saw back and forth on the grate of the Night Sun. That's a hell of a sight, all right, Kalt thought. If that cable snaps or bird-cages, it's over.

Below them, riding up and down the crests of the waves, was a man in a survival suit. He was floating, spread-eagled and facedown, but neither his arms nor legs seemed to be moving. He did not know if the man was alive or not, but he certainly was not moving on his own.

A hundred yards off was the strobe light and a jumble of retrotape.

Better go for the ones who look like they're conscious, Kalt said to himself. Get moving.

He conned Torpey to fly the helicopter far beyond the right of the strobe before putting the basket out. Then he instructed the pilots to move gradually back toward the left, dragging the basket through the water as they went. He and Torpey understood each other perfectly now. He only had to call two or three conning commands to establish a hover position over the strobe.

They were fifteen minutes into the hoist evolution when Ted LeFeuvre noticed the warning light flashing on the fuel gauge.

"Steve," he said, "we've just exceeded our BINGO."

"Oh."

"We don't have enough gas to get back to Sitka."

Torpey did not answer him. He was banking the helicopter and fighting to hold a position.

"I'll figure it out."

He could sense that his brain was slowing down; to calculate a simple fuel-burn rate took Ted LeFeuvre more than a minute. As he figured how much gas they would need to reach shore, he was thinking: We can't just leave these people here. I didn't ask to be here. But we've gotten this far. If we stop now the rest of those fishermen are going to be fish food.

Since he needed both hands to work the collective, he used the foot switch to make a radio call to the C-130. The plane had arrived on scene just as the second helicopter was aborting mission and returning to Sitka. He had not yet spoken to the C-130; Fish had been working radios. But when Ted LeFeuvre heard the unperturbed voice of the radioman, he felt a renewed sense of calm come into him. Confidence, like fear, was contagious, and now the confidence he had before the first rogue wave had nearly finished them was back.

He identified himself and told the radioman that their flight ops were normal.

"*Roger, 6011,*" the voice said. "*How are things down there?*"

"Rough."

"*How bad are the seas?*"

"Bad."

"*What can we do for you?*"

"Listen," Ted LeFeuvre said, "I'm pretty worn out and I wondered if you guys could confirm a fuel-burn rate for me. We're looking to go to Yakutat. Now as I see it, that's sixty miles north and east, so with a tailwind of seventy-five knots, I'm figuring it will take us fifteen minutes to get there."

There was a silence.

"Is that what you get?"

"*Hold on.*"

He waited twenty seconds, each feeling like a full minute, and then he heard the radioman say: "*Seventeen minutes flight time is what we get from your present position, over.*"

"Thanks," Ted LeFeuvre said. "We'll talk in fifteen minutes. Rescue 6011 out."

He turned to Torpey.

"Listen, from here Yakutat is about fifteen minutes, which means we've got enough fuel to safely stay for another hour and forty minutes."

"Are you sure?"

"I just double-checked my figures with the C-130. They came up with the same thing."

"Captain," Torpey said, pointing, "watch that wave there!"

Ted LeFeuvre hit the collective, heard the turbines whine and felt the sudden, hollowing-out thrusting jump of the helicopter in his stomach. A comber—eighty feet at least—swept beneath them. Torpey exhaled.

"Okay," he said. "We stay longer."

Below them, the wave buried the basket for almost a minute. But Kalt did not stop dragging it until it was within ten yards of the strobe light.

"Paying out slack," he said.

Ted LeFeuvre tried to ease the helicopter a little lower, to eighty feet now, to give Kalt an extra twenty or thirty feet of slack cable. With a lot of slack, he figured, there was less pressure on the line, less chance of damaging the winch and the cable itself.

He was dropping their altitude when he heard Kalt shout: "*Survivor's in the basket!*"

Just then a gust buffeted the helicopter.

As he pulled lift power he heard the winch screech and the hoist cable lash the airframe. Honnold, Fish and Kalt were shouting. The hoist was screeching. Torpey was holding the cyclic to the dash and yelling something he could not hear. It would not have helped him if he had.

Kalt struggled to the winch, found it in the stop position. The cable was jerking and more than eighty feet of hoist cable were still out.

"I'm pulling it up," he shouted to Honnold.

He shoved the hoist in gear and, with one hand on the grab rail, leaned halfway out the helicopter. The hoist was still spooling smoothly.

Then *wham*—the helicopter was over on one side and he was skidding on the deck.

He struggled to his knees, checked his helmet. He was all right. He stood up in a crouch. Lousy, bitching gusts, he said to himself. He looked down out at the raging sleet beneath the helicopter, the flakes long and white as chalk in the floods.

"Hey," Kalt said. He sounded as though he could hardly believe what he was saying. "Someone's still in the basket."

"Move your ass!"

"I am."

"I said move it!"

"I am moving!"

"You want me to leave you behind?"

"No!"

"Then *swim*, you fuck!"

Ahead now they could see the green glow of the chemical lights, appearing and vanishing behind the swells. Otherwise the spray and sleet were so thick they could hardly pick out the waves.

"Swim!" Gig Mork shouted.

"I can't!"

"You lousy cunt! Swim! Swim!"

"I'm trying!"

"Harder!"

Mork was holding Mike DeCapua with one arm and flailing and swimming with the other, and it was as though they were moving uphill and downhill, not sideways, through the breakers. He looked up and the green box was coming closer and he thrashed and fought through the water, the spray clawing at his eyes, and he kept thrashing and swimming even though his lungs felt as though someone had thrust a hot poker through them, the pain from not breathing so great, and everything was turning black and his throat filling with ice water when he felt the hoist baskct in his grip.

"Hold this!"

While DeCapua steadied the bobbing cage, Mork grabbed the crossbar and hoisted himself up into the basket.

"Get in!"

The basket slipped right out of DeCapua's hands. He fell backward. The EPIRB was gone.

Mork had him by the legs.

"I got you!"

He pulled DeCapua on top of himself. With the extra weight and the cable slack, the basket sank.

"Get your leg out of my face!"

Just then a wave toppled down on them like a wall of bricks and the next thing Mork knew he was one leg out of the basket, one foot on the top of the cage, his hand barely holding the cable. The basket was twirling like a slowing top, scudding foam and spray as it twirled, and he knew he was going up. He was going up fast and all he knew was the flying ice and black and the cable, and all he could do was squeeze the cable with his death grip. Don't let go of this thing. You do and you're dead. Christ Almighty, speed this son of a bitch up.

The first thing he saw was the door and then he saw a huge man wearing a shiny, black helmet. He was beautiful. Then a big glove reached out and seized the cage and then a second glove was seizing him by the shoulder and he was inside the cabin.

He was lying on the deck alongside two black boots. He coughed out seawater and rolled over on his back. His knees and elbows hurt.

Only then did Gig Mork realize that he had come up in the basket alone.

F O R T Y - E I G H T

The basket was going down again. The wind had nearly swept it into the tail rotor after Fred Kalt pitched it out the jump door and he was now hurrying to get it down to the third survivor. Suddenly he felt the line jerk.

He cut the power on the hoist motor and line stopped going out.

"Mr. Torpey," Kalt said, "I think we've got a broken strand in the cable."

"No."

"Yeah."

Throwing the winch in slow reverse, Kalt knelt down and let the cable slide across the palm of the thick, leather glove used for hoisting.

He frowned.

"What is it?" Lee Honnold asked him.

"We got a burr."

"Where?"

"I don't know. I thought I felt something right around here. How much more line we got out?"

"About forty feet."

"Damn."

Kalt kept feeling the line to find the nick. It was dark and hard to see. The pitching and jumping around did not help. The Coast Guard used three-sixteenth-inch cable for hoisting. It was 105 strands of stainless steel, tightly woven, and each cable had a breaking strength of eighteen hundred pounds. But once it had a kink or a burr, you were working on borrowed time.

It won't take long now, Kalt was thinking. All it takes is one strand to go. Just one little kink. With the tension on it and the beating this thing is taking it'll be nothing for the whole damned thing to start unraveling.

"You find it?"

"Not yet," Kalt said.

You can do a quick splice, he was thinking. You'll have to cut the cable where the strand broke, do a splice, resit the hook. There's 200 feet of hoist cable. Okay, so if you cut off 40 feet, that'll leave you 160 to hoist with. Is that enough? Not if we're hoisting from a height between 100 and 140 feet. We won't have much left for slack. You need that extra slack to keep the basket from moving around too much in the waves.

Steve Torpey's voice crackled over the intercom. "How bad is it, Fred?"

"I don't know, sir," Kalt said. He was still looking for the kink in the cable. "I'm sure there's a rough spot, sir. But I can't find it. Do you want to continue?"

We can keep going for a while, he was thinking. But what if we get that third guy in the basket and start hauling and the line birdcages? Then what? It's going to be just like when a shoelace starts fraying and you try to pull a strand of it through a hole in your shoe and the rest of it bunches up on one side. The cable is going to bunch up in the guide chute to the drum. Then the reel is going to jam. And then we're going to have a survivor swinging in the basket forty feet below the helicopter. Then what do we do? Pull him up by hand?

"Mr. Torpey?"

Torpey had been thinking hard. He was finding it more difficult now to concentrate. "Listen, Fred," he said, after a long delay, "let's just keep going. If we don't do something, the man's dead anyway."

"Roger."

Kalt threw the hoist in reverse again.

"Okay," he said, "line is going out." He looked down at the churning sea and saw a splash.

He said, "Basket's in the water."

Ted LeFeuvre was keeping a close eye on the gas gauge. They had less than forty minutes of fuel left. We've got enough for another four, perhaps five basket drops. No more. After that, there'll be nothing to do but leave whoever is down there to the grace of God.

Down in the sea, Mike DeCapua was just about out of his head. He had not been able to feel anything in his hands and legs for quite some time, and his feet, as far as he could tell, were as good as gone.

He could hear the helicopter, the dull thudding of the rotors mostly, but he had lost sight of it. Some of the flares were still burning. He could see them when a big swell lifted him up above the other waves. But he knew that soon all of the flares would go dark and he remembered he no longer had the EPIRB. That was long gone. No EPIRB, no strobe, no way to find Mike, he thought. Then those flares are going to burn out. Then it's game over.

I don't want to give up, he thought, but what choice have I got? No hands, no legs, no feet. I must have the hypothermia. That's what I got. Christ. If they were to pin me up against a wall and tell me the firing squad would let loose on me if I didn't stay up on my feet, then I guess I'd just have to shut my eyes and wait for it. It's a weird thing, this hypothermia. You don't feel your nuts. But you can feel your body temperature dropping. I wonder how that is? Maybe it's because you've stopped shivering, he thought. But that was a while ago. Still, maybe that's it. Maybe that's how you know. I wonder. Well, any way you cut it, it's weird.

Everybody in Alaska knows about the hypothermia, he was thinking. One of the first things you learn. One of the first things you put out of your head. What is it they say? After you quit shivering you got no more calories to burn. Calories. That's where you get your heat from. I guess fat people are pretty hot. That's funny. That's real funny. A fat, hot woman. Sure. Only in America. Do you suppose I lost my fuel supply? I guess so. My extremities have already shut down. I must be experiencing the numbness that happens when your body quits pumping blood to your extremities. That's what I'm experiencing. It's the natural thing, I guess. It's the way nature must work.

Gee, you're real smart, Mike. Real fucking smart. You always did have a good theory to explain things after the fact.

I'm tired, he said to himself. Whipped. I wonder if it would do any harm to sleep? Just sleep. That's all I want to do now. It wouldn't be that hard to do. It's never hard to sleep once you're beaten. Am I beaten? I'm beat. That much I am. I'll bet if I closed my eyes I'd go right off. Ain't nobody out here to tell me what to do. It'd be real easy, all right. Just close my eyes and slip right off the edge.

I wonder if this is what Hanlon was feeling when he went under? Or was he hot? They say some guys get real warm at the end. Nice and toasty. Shit. You think the hypothermia would make me feel toasty. How do you suppose he came out of that knot? I tied that cat's-paw as good as it could be tied. And it's a damned good knot. It don't come undone by itself. If you back it up with a half-inch, it won't. But it's an easy knot to break. What do you suppose he did? Maybe he panicked. Got himself in a panic and tried to breathe underwater. Or maybe he said fuck it and just swam off. I don't know. I tied that buoy ball to him, too. Just like I tied this one to me. Right around my middle. Where is it now? Right around my leg. It must have slid down some. Great. Fucking buoy did me a shitload of good. Fucking buoy ball.

I should have stayed in that fucking basket. If I hadn't fallen out I'd be up in that helicopter now. I guess Mark and Giggy and Bob are sitting up there right now laughing their asses off and drinking something warm.

Well, maybe they ain't laughing. Christ, that was a bitch of a wave. Knocked me right out of that damned basket. That was some fucking fall, that was. I wonder how far I fell? Twenty feet? Thirty feet? Good thing I'm too cold to feel it. Real good thing. I wonder if I broke anything? Nah. You'd know it if you did. Or would you? All I wanted was to get in that basket. That's all I wanted. Well, I'm going to sleep now. I'm giving up. I want to go to sleep.

Then his mind started jumping around and he felt something passing through him like waves and then he was crying. He cried for a while, and then after a while it was all right and he started to think about his daughters, Misty Dawn and Melanie. What a stupid thing, he thought, letting nineteen fucking years go by without seeing your kids. I wonder what they look like now. Nineteen years. That was a lifetime to some people. And Mario. Signing those custody release papers on Mario was not one of my better moves. No, sir. All I ever wanted was Mario and Robin. Not her two lousy kids. Well, you made the deal. You signed the papers. I hope Mario's okay, he thought. I hope Mario is okay and I hope he has a good life. Too bad I was such a prick. He didn't deserve that. He deserved a dad. Well, I guess he's got a dad. I guess he's got a good one now. Sorry, kid. Sorry, girls. I guess we all had to live with it. I never used to realize that, I guess. I just played it along and played it along and never cared about the consequences. I knew about them but just never cared. Well, it looks like I ain't living with any of it anymore. I guess I find out now whether there's a heaven or a hell. I'm tired. I'm so fucking tired. The hell with this. I'm going to sleep.

He was just going to curl up into a ball when he saw the rescue basket.

It was riding a swell, a big swell, and at first he thought it was a mirage, a hallucination. It was all lit up, a bright, starry green, sparkling, like a Christmas tree. Then he remembered the glow sticks. There had been glow sticks on the basket. That's no mirage, he thought. That's the real thing. That's a rescue basket.

Jesus.

And he was moving toward it. He did not understand how. His feet were not working. His hands were not working. Yet he was moving toward the basket. He did not know whether he was swimming or not. It did not feel like he was swimming. He was not asking questions. He was just happily going toward the basket. Or was the basket coming toward him? He did not know. He did not want to know. He and the basket were getting closer and closer together and everything went calm around him, everything, the waves and the wind and the snow and sleet and spray, and there was a big pause, sort of like a missed breath, like a rest in music, and the waves did not seem as big as before, and he was happy and not asking questions, just saying *Thank you, Thank you,* and the next thing he knew he was inside the basket and breaking free of the water and something was whispering to him: *This is your miracle.*

He was clear of the water and rising toward heaven and feeling relief, the lightest, wildest, most unearthly, immense spasm of relief he had ever felt and then Mike DeCapua was in the helicopter and someone was tugging on his legs.

"That's the last one," he heard a voice say.

His head flopped on the deck to one side. He saw someone in a survival suit. Then a knife. Someone was leaning over him with a knife.

"Don't move."

"Thank you . . . thank you . . ."

"Don't move."

"I lost it . . . the EPIRB."

"Lie still," the voice said gently. It came from a helmet. A man in a suit was attached to the helmet. He could see him holding a little knife.

"Please," DeCapua said, trying to shake his head. "Use the zipper. Don't cut my suit."

The knife was doing something and then it went away. Hands were tugging on the shoulders of his suit.

"I . . . I can't . . . can't get up."

"Lie still."

He was shaking so hard now that everything in the cabin looked blurry.

"How are you feeling?"

He did not know the voice. It was like a kid's voice, clear and pleasant sounding. Giddylike. Someone's hands were stripping him of his suit.

"Cold," he said. "So . . . cold . . ."

"I see that," the voice said. "You were in the water too long. Didn't you know you shouldn't be swimming this time of year? Okay, I'm going to slip a thermal capsule over you." While the voice went on talking he felt the hands pulling on his suit. Then DeCapua felt something plastic around him. "That's it. How does that feel in a capsule?"

"Not bad."

"Can I get you anything?"

"Cigarette?"

"Well," the voice said, "that won't happen for a while."

DeCapua closed his eyes. The shakes were coming worse now. They felt good. He could feel a little spot on the small of his back warming.

He turned his head.

"Where's Mark?"

Fred Kalt was the first to hear it: a series of dull, clubbing thuds coming through the back wall of the cabin, just aft of the jump door.

He and Lee Honnold leaned out for a look.

A rope had wrapped itself around the black, metal pipe that stuck out near the tail of the aircraft—the high-frequency antenna. On the end of the rope was a buoy ball.

The float was flailing in the wind and bouncing on the stabilator, the fin that kept the tail of the aircraft level. It was hitting no more than three feet from the tail rotor.

"Holy shit."

"This is *nuts!*"

Kalt had seen the buoy ball loosely tethered to the leg of the third survivor as they were hoisting him. As soon as they had pulled the fisherman into the cabin, he had severed the rope. The buoy ball, still dangling outside the helicopter, had apparently sailed straight back.

Oh, Christ, Kalt thought. He was feeling the strength drain out of him in a steady, faint nausea.

The taillights shone on the bright yellow ball as it *thunk-thunk-thunk-thunked* on the stabilator.

We're in some spot, Kalt was thinking. Some spot. Had that rope not caught on the antenna the way it did, that ball would have flown right into the rotor. And that would have done it. That would have done it for sure.

"What do we do?" Honnold asked him.

Kalt gripped the railing and leaned heavily against the doorjamb, unable to take his eyes off the rapping float.

Forget about the rest of it, he was thinking. Forget all of the other close calls. And forget about getting the survivors. It's all right there in that ball. If that float hits that rotor, we're in the water. We're in the water in fifteen seconds.

"Fred?"

Just then they heard a steely yawn and the antenna snapped off its housing and went twirling off. The buoy ball and rope shot straight up and planed off into the darkness.

Kalt lowered his head and muttered. Honnold rested a hand on his shoulder.

"Cabin crew?"

That was the captain's voice coming over the intercom they heard. They did not answer him right away.

"Fred?" Ted LeFeuvre asked. "Do you hear me? Fred? How are we doing back there?"

"Fine, sir," Kalt responded in a little voice. "Everything is just fine."

The flares were fading now, going out.

"Fred, get some more of those Mark-58s ready," Steve Torpey said to Fred Kalt. "I'm not going to try to hoist that last guy without them."

"How many?"

"Six," Torpey said. "Get six up. But this time I'm going to fly a racetrack pattern when we drop, okay? We're going to fly over the skipper. Then we're going to turn, circle around and see if we can't scoop him. Understood?"

"Roger," Kalt said.

Torpey turned to Ted LeFeuvre. "How much fuel have we got, sir?"

"A half hour. A little less."

Torpey nodded.

"Let me know when, Fred," he said to Kalt.

"Two minutes."

There's really nothing else we can do, Ted LeFeuvre was thinking. The skipper apparently had not moved since falling from the basket. But what if the man was just unconscious? What if he was awake but could not move? We've got to make a try. These young guys never get tired. I'm tired. But we've got to try. Keep your eyes out for that fifth fisherman, too. He may be down there. Sure he's down there. Where exactly, I don't know. But he's down there. Somewhere.

Mike Fish spoke up. "Sir, I know what you said earlier, but I could go down and get that guy in the basket real quick. I'm ready to go."

"No, Mike," Torpey said firmly. "No."

"Yes, sir."

Fish sat back. Gig Mork, who was sitting alongside him against the back wall, tapped him on the shoulder. The rescue swimmer leaned over.

"Yeah?"

"Listen," Mork said, "if it's all right with you, I'd just as soon go back down there myself and get him."

"Oh, no," Fish said. "Nobody's going out. You just sit tight. We're going to try to pick him up now. But nobody's jumping back out in this."

Mork looked at Fish and nodded.

They dropped the smokes in a fifty-foot diameter around the survivor and Torpey banked them in a three o'clock hover. Kalt added a weight bag to keep the basket from sailing and splashed it down not five yards from Mark Morley, but the skipper did not seem to move. They winched the basket up, tossed it out again, and this time dragged the basket until it bumped him. Kalt did this three more times. On the sixth drop the basket landed smack on Morley's outstretched legs. But he made no moves for it.

Kalt even tried scooping him up by dragging the basket through the seas, but that didn't work either. As he winched the basket up to try again, Torpey, who had said very little except to respond to Kalt's conning commands, said: "Any change?"

Kalt said, "He's been floating facedown, spread-eagle, the whole time."

"No movement at all?"

Kalt was looking out the door. "No," he said. "None. No movement, sir."

Ted LeFeuvre had been following their progress through the chin bubble on the cockpit floor. For the most part he had watched, silently, as they dropped and hoisted. He felt Torpey and Kalt had developed a rhythm, an understanding during the first three hoists, and he did not want to break it up. Even then they were working well together, and it was a great thing to watch. But he could see Torpey was tiring. The same difficult maneuvers the young pilot had performed with such grace and sharpness earlier were no longer as crisp. He's not quite erratic yet, but he's getting pretty close, he said to himself.

"How're you doing?" he asked Torpey.

"Fine."

Ted LeFeuvre checked the fuel gauge. They were less than ten minutes from the BINGO limit.

"Hey, Steve," Ted LeFeuvre said, speaking quietly, "it's time for us to go."

"Captain?"

"No, Steve. I said it's time to go."

"No, wait, wait," he heard Kalt say over the intercom. "I almost got him that time. We can do this. We can do this, sir!"

"No, Mr. Kalt." He said it with all the efficiency, coldness and swiftness of a commander who was implementing a strategy that was beyond discussion.

"Captain," Torpey said to him, "we can get this guy."

"No," Ted LeFeuvre said. "It's time to go. *Now.*"

Torpey looked at the snow sweeping across on the windscreen. Fish sat reading the radar altimeter. Lee Honnold clicked off the handheld searchlight without a word.

Kalt looked at them all, and nodded.

Ted LeFeuvre said: "Prepare to depart." There was a silence. "Begin the third rescue checklist."

"Yes, sir."

"Roger, Captain."

"Captain," Torpey asked, "maybe we should dump some more flares around the skipper so that the next aircraft will more easily spot him."

"That's fine."

"Steve," Ted LeFeuvre said, "why don't I take the controls and fly us to Yakutat?"

"That sounds very good to me."

"You handle the radio and the flight computer."

"Fine."

As Ted LeFeuvre pulled power and the Jayhawk climbed to three hundred feet, Fish and Honnold opened one last batch of flares and handed them to Kalt, who armed and swept them out the cabin door.

Ted LeFeuvre was mapping things out as carefully and deliberately as

he could. Yakutat was about fifteen minutes of flight time away. Let the
tailwind do the work, he said to himself. Sixty miles. That's a snap. Just
don't keep the aircraft too sharply out of trim. Fifteen minutes. Maybe it
was sixteen minutes. Gee, I'm tired. Running on fumes. Bank the helicopter. That's it. Now drop the nose. Turn us slowly. Nice and easy. Okay,
now, let's go to six hundred feet. That'll put us right below the freezing
level. Oh, my. Feel that tailwind. Wow. We're doing 210 knots. Let's go,
LeFeuvre. Stop watching and start flying. We got three of them. We lost
the one. But we got three. No, that's not right at all. We got more than
three. We got eight. Eight of us are going back. And don't start with the
whys. Don't start that. There are too many whys in this business. Save it for
some other time. Just get this plane to Yakutat. Just get us there.

Behind him in the cabin, Kalt was clearing the deck. Honnold and
Fish were shaking hands and chatting with Gig Mork and Bob Doyle.
They gave the survivors some hot tea and water and then took their seats
and strapped themselves in. Mike DeCapua was stretched out on the floor,
quietly shaking.

Mork looked at Bob Doyle.

"You need a beer."

Bob Doyle tilted his head to one side and said nothing. The indigo
cabin lights went out. He closed his eyes.

Fish asked: "Either you guys want more tea or some coffee?"

"Sure," Bob Doyle said.

Fish handed him a thermos cup of hot tea, and Bob Doyle took a short
sip. The tea warmed his throat. It didn't warm the rest of him. That'll take
some time, he thought. That'll take a lot more time.

Kalt slid the door shut and bolted the air lock. He pulled off his helmet
and hoist gloves. His face was white and so were his hands. The muscles in
his legs fluttered and a queasiness rose in the pit of his stomach. He sat
down on the deck. Then he reached for his flight bag, opened it and rummaged around until he found an orange. It was hard to peel the orange.
His hand kept shaking.

At his feet sat the rescue basket. The light sticks attached to it still
glowed a feeble green. Kalt stared at the sticks until the glow in them
died out.

BOOK SIX

They flew through the dark, keeping the wind at their tail. The sleet and rain had let up and only came occasionally in gusts. Once they made the shoreline they swung north and followed the coast toward Yakutat. They had hoped to see some sign of life below, a light, a ship, breakers on rocks, perhaps. But there was nothing. The ceiling had come up above fifteen hundred feet but it was very dark and the wind was blowing a gale. Steve Torpey turned up the high-frequency radio and started making calls in the blind. He told the world that they were about fifteen miles to the south of Yakutat and preparing their final approach for landing. He advised all aircraft in the vicinity to take note of their course. There was no response. Yakutat was not a tower-controlled field. It was nearly four o'clock in the morning. Nobody would be out there listening. He turned down the radio and sat quietly in his seat.

For several more minutes they saw no lights, nor did they see the shore, but they flew steadily in the dark riding the tailwind. It was quite rough. Ted LeFeuvre kept the same speed until he saw rocks rise up from below, the waves striking against them, rushing up, falling back. He throttled down the engines a bit and began making a slow turn into the wind. The helicopter shuddered when he did it. They were back out over the ocean. Then he lined them up with the airstrip. The Yakutat airstrip sat two miles inland and was short and straight, running west to east. Ted LeFeuvre wanted to land with the tailwind. The wind was still blowing strong.

"There it is," he said.

Torpey clicked on the nose lamp and the belly floods and they could see the wind driving the spruce so that the tops shook and swayed and shed bows of white, looping snow. There was no moon but it seemed much lighter than it had been before, and they swooped in over the tops of the spruce jaggedly reaching and swaying, and dropping the helicopter on a steady angle now, lining up with the blacktop, Ted LeFeuvre pulled the nose up just a touch and the trees went slipping smoothly by and then there was the familiar, heart-tugging catch of the wheels on the asphalt and they were bumping quickly along the runway. He slowed the Jayhawk into a trot, finishing off his running landing, and then taxied them over to the hangar. He turned the aircraft so that the nose pointed into the wind and then he shut it down.

The rotor head slowed, then stopped. The blades were full of wind and flapped up and down and smacked the pavement. He sat back in his seat. He was very, very tired of flying. His arms and shoulders and back ached and his hands were stiff. He looked at the airspeed indicator. They were on the ground and the aircraft was stationary. Yet the indicator was still showing sixty knots. Ted LeFeuvre looked at the number for a moment and loosened his chinstrap.

"Nice landing," Torpey told him. Ted LeFeuvre pulled off his helmet. "Thanks."

He tried the door. It felt heavy. He pushed harder and it opened a crack.

Finally he leaned his shoulder into it and this time the door opened and he stepped down on the wet pavement and was in Yakutat. The tarmac felt very pleasant under his boots. He walked around the nose of the helicopter. It was pleasant to walk on firm ground. There were a lot of people milling about. There was an ambulance and two police vans and some cars and a Dodge pickup truck. Someone must have telephoned the airport manager, Mike Hill, he thought. District 17 must have phoned ahead of their arrival. He stopped and shook hands with Hill, and then he saw Kip Fanning from the Yakutat Lodge and Les Hartley of Gulf Air airlines, and a few other people whose names he had forgotten and was too tired to remember, and he saw the cabin door open. Lee Honnold and Fred Kalt jumped out and patted each other on the chest and then gave each other a bear hug. Then a man in survival gear, short and bleary and shaky, stepped down from the aircraft. He shooed away the medics. He didn't want their help. The man walked unsteadily, side to side but with a rough dignity, and then climbed up into the ambulance.

Then he saw Bob Doyle.

It was not the thick, scraggly beard, the long hair or the gray, drawn face that made him watch. It was something else. The man was helping a medic with a stretcher. He was helping to carry the shivering survivor, whispering to him as they went, and then Bob Doyle lifted the sagging stretcher up and eased it into the ambulance before climbing in himself. He did not bother to look back. The doors shut and the ambulance pulled away.

Ted LeFeuvre stood watching it.

Torpey walked over and stopped beside Ted LeFeuvre. He followed the captain's gaze. The ambulance was now turning out along a connector road.

"Unbelievable," Torpey said. "Unbelievable."

The big trees at the end of the runway were swaying far over in the wind. Torpey scratched his head.

"It's just so hard to believe."

Ted LeFeuvre did not hear him. He was watching the ambulance get smaller and smaller and thinking. After all he's been through tonight, he said to himself, Bob Doyle had every right to lie down on a stretcher himself and do nothing but take a ride to the hospital. But instead, he thought first about his buddy. He helped his buddy into the ambulance first. It was just a little thing. But pretty noble. That was noble of him, all right.

He sighed.

"What's that, Steve?"

"I said I just can't believe it."

"Yeah," Ted LeFeuvre answered, "I can't believe it myself, Steve. I mean, those were the worst conditions I have ever seen. Those waves and that wind—"

"No, no," Torpey interrupted him. "I wasn't talking about the weather."

"No?"

"No."

"Then what?"

Torpey looked at him.

"Well, sir, for a whole year I looked high and low for Bob Doyle at the air station and could never find him at his desk. Not one time. Now I fly a hundred and fifty miles offshore at night, in the dead of winter, in a hurricane, and there he is, floating in the Pacific."

He shook his head.

Ted LeFeuvre started laughing and could not stop.

———————

At daybreak it was still blowing thirty-five knots, but the seas were down to forty feet and the sleet had softened to a cold rain that danced along the swells.

Two C-130 planes and an H-60 helicopter from Kodiak took off for the Fairweather Grounds while the cutters *Anacapa* and *Planetree* steamed out to assist in the search for the two missing crewmen. By 9:40 A.M., a data marker buoy had been dropped at the survivors' last reported position, 58°22' north, 138°42' west; the radio float began drifting in the counter-clockwise current of the gulf, and the search area shifted continuously north and westward at about two knots an hour.

The aircraft crisscrossed the grounds for hours, flying sector searches in an eight-nautical-mile radius, then in twenty-five-mile rectangular grids. But with the rough seas, fog and steady rain, they turned up nothing.

A little after one o'clock, the oil tanker *Arco Juneau* arrived on scene. It had been transiting south from Cordova Bay the previous night when its captain, Mike Devins, decided to veer north and west of the Fairweather Grounds and ride out the blow in deep water. Throughout the night, the tanker slugged its way south and east—120 miles—and, at 1:55 P.M., reported seeing an orange object riding the swells, which it later described as looking like somebody in an orange survival suit. The news was not encouraging. The survivor was floating on his back, arms and legs spread wide, and showed no signs of movement. Despite high, heaving seas, the *Arco Juneau* promised to keep the survivor in its lee.

Within thirty-six minutes, an H-60 sent the previous night from Kodiak to Yakutat was on scene and lowering a rescue swimmer in a lift basket.

It took four minutes to haul up a man in an orange survival suit.

With his suit full of seawater he weighed close to four hundred pounds. He needed a shave. His open eyes were bloodshot and shiny and almost seemed to hold a self-satisfied expression. His mouth was open a little, as they usually were, and showed big, strong teeth.

The flight mechanic gently laid the head on the deck, then slit the survival suit. He rummaged in the man's pockets and found a wallet.

Stuck to the back of a clammy, wallet-size portrait of Jesus Christ was a photo of a young woman with wide-set, green eyes and a laughing mouth.

Cute, the flight mechanic thought. Too bad for her. Too bad for him, too.

Less than a half hour later, a nurse and a medic wheeled Mark Morley through a corridor in the Yakutat Hospital and into the emergency room. Three fishermen, two of whom had just been released after treatment for hypothermia, were smoking in the lounge when the gurney was wheeled in. They watched the door swing shut.

A short while later the nurse came out.

"How is he?" Bob Doyle asked.

She shook her head.

"Oh."

"He's dead," the nurse said. "If you want to go in now and see him, you can."

Gig Mork said nothing. Some townsfolk had given them some clothes and donated sneakers. Mork just stood there in his new sneakers, biting his lower lip.

Mike DeCapua said, "No, not me, thanks."

"I'd like to see him," Bob Doyle said.

"In here," the nurse said.

They walked into a white room where Mark Morley lay on a wheeled table, a sheet over his great body. The fluorescent light left no shadows. Under the harsh light his inflated face looked freshly shiny. The scalp was lacquered red around the gashes. The eyes had that glassy, flat look. Bob Doyle might as well have been looking at a couple of bottle tops.

His throat had swelled up so it was hard for him to talk.

"Take your time, Mr. Doyle," the nurse said. She put a hand on his shoulder. "I'm very, very sorry." Bob Doyle did not seem to hear her.

"Oh, Christ," he said, and he began to cry. "I should have jumped in after you."

They left the hospital that afternoon. Bob Doyle and Mork had suffered only mild cases of hypothermia. DeCapua, who looked pretty bad when he arrived at intensive care, recovered quickly. He hadn't suffered any nerve damage or circulatory problems as a result of his hypothermia and he wouldn't stop bumming cigarettes off the nurses, so the doctor released him that afternoon.

The three of them were flown down to Juneau that same night. Scott

Echols, who had owned the *La Conte*, met them at the airport. He drove them straight to the Best Western hotel downtown and paid the bill. He ordered out a pizza and put a hundred-dollar check in each of their hands, along with a one-way plane ticket to Sitka.

"I'll take care of you guys," Echols was saying. "I'll take good care of you. I owe you guys a lot."

Bob Doyle looked at him.

"The important thing is that you're alive," Echols went on. "I want you to know that we're going to keep in touch, and that if a job ever comes open on a boat of mine, for a fisherman, you guys will be on the top of my list. That's a promise."

They stared blankly at him.

"Don't you worry about a thing," Echols said.

"I'm *not*," Bob Doyle said.

"Well, good."

Mike DeCapua said to Echols, "Say, I got something I got to square with you."

"What's that?"

"It's about the suit. The survival suit. That suit—the one Hanlon was in—see, I borrowed that suit from a buddy of mine off another boat, see. I got to give him back something."

"Oh."

"What I'm saying is, if you don't mind, I'd like to keep my survival suit. The *La Conte* suit. The one I still got. To replace the one that Dave is wearing—was wearing."

"All right."

"I'm sorry. I know you already lost a bunch, with the boat and all. It's just that, well, it wasn't mine."

"Sure," Echols told him. "That's okay, Mike. Go ahead. Take the suit."

"Yeah?"

"It's not a problem. It's just lost equipment. Don't worry about it."

Bob Doyle was watching him. Echols smiled. "You know, Bob, I had no idea. No idea at all. He should never have taken you guys out there."

Bob Doyle looked at him.

"Look," Echols told him. "I just want to help you guys out. I feel like I owe you guys."

"Oh yeah?"

"Listen, Bob. Anything you need. Anything. You just let me know, now, all right?"

Within a month, Scott Echols had shut down World Seafood Producers, moved out of his duplex in Juneau and relocated with his wife to Guam.

F I F T Y

It was close to six in the morning when Ted LeFeuvre heard the C-130 plane landing outside. He had been lying in a hard, wood bunk at the Yakutat Lodge with his eyes shut and his mind jumping around. He went to the window and looked out. It was still dark and a light rain was falling. There were glistening, moving patches that the landing lights made across the blacktop and he heard the drone of propellers and voices. He pulled on his flight suit and boots, brushed his teeth and combed his hair and walked down a long corridor to the dining hall.

He sat in one of the heavy wood chairs and waited for a server to come. No one came out so he went to the counter and got himself a menu and poured himself a glass of water from a pitcher and sat back down and glanced through the menu. After a while a woman with bleary eyes and a smudged apron came over and he asked for hot rolls and scrambled eggs and juice. She went out and he drank the water. He saw the water jump a little in the glass and noticed his hand was still a little shaky. He put the glass down and Lee Honnold, Fred Kalt and Mike Fish came in. They sat down.

"How did you sleep, Captain?" Kalt asked him. He had those dark bags under his eyes but he looked alert.

"Fine."

"The C-130 is here."

"Yes, it is."

"Have they sent over any helos from Kodiak?"

"I think so."

"Well, if they haven't, they probably will in no time," Honnold said.

The four of them sat at the table and ate and then Steve Torpey came out and had coffee and eggs and more coffee. Fred Kalt and Lee Honnold ate fairly quietly. Ted LeFeuvre watched them as he drank his juice.

He had heard them talking in their bunks, after lights-out. The lights had gone out and then Kalt had said something about how they had let the skipper fall. Honnold had wondered why they had not seen the guy dangling from the basket. I don't know, Kalt had said. I don't know how we missed that. But we sure missed him all right, Honnold had said, we should have seen him. We should have. There was no excuse for that. No, Kalt had agreed. We should have gotten four, not three survivors. How could we have missed him?

"By the way, you guys," Ted LeFeuvre said, looking at Kalt and then at Honnold as they ate their breakfast, "I don't want to hear any more talk like I did before, at lights-out."

"Sir?"

"You know what I'm talking about," Ted LeFeuvre said. "All that business about that guy falling. That's no way to talk. I don't want to hear anyone talking that way."

Kalt and Honnold looked down at their plates.

"That was no way to talk," the captain said. "Remember, three men are going home to their families because of you guys. That's three happy families. Three. So I don't want to hear any more about who we lost. No more about that. Is that understood?"

They nodded and kept on eating.

When they finished breakfast, Ted LeFeuvre paid the bill and they walked outside to look over the aircraft. A fresh H-60 crew had just deplaned from the C-130. They were doing the maintenance checks on another Jayhawk, the 6041. The 6041, they learned, had flown in from Kodiak during that hour or two that they had lain down to rest. The new crew was going out to the Fairweather Grounds to look for the skipper and the other missing fisherman. They looked sharp and eager. There was a certain quickness to their step that seemed foreign to Ted LeFeuvre. All he wanted was to get home and sleep. It was a bad thing not getting any sleep. Made you nervy. He was actually looking forward to seeing his bed. That was something. That was different.

For several hours Kalt and Honnold inspected the 6011. The hoist cable was shot. The winch would need an overhaul. There were scratch marks and gouges on the helicopter's belly and side. Ted LeFeuvre noticed the same gouges on Honnold's helmet, too, but he did not say anything. The Night Sun was bent. The engine had been overtorqued. They ran some tests on the transmission. It seemed all right, although the auxiliary power unit had broken. It took Honnold most of the morning to repair it and power up the aircraft. They didn't get airborne until almost two that afternoon.

Normally they would have saved some time and fuel and flown a straight line south across the gulf to Sitka. This time they hugged the coast. Nobody said anything during the flight. After an hour, Mount Edgecumbe was below them. Its base looked washed and black and there were clouds swathing its flat, white top. A fog was coming over the mountains from the sea. Beyond lay the channel and more mountains and the town, fuzzy in the gray rain.

As soon as they touched down all of the enlisted personnel at the air station, the mechanics and machinists and shopkeepers and electronics techs, the ones who had worked all night, came walking out on the tarmac. It was drizzling and they all stood outside in the rain and waited for them to step down from the helicopter. Ted LeFeuvre nodded to them and smiled. He felt very, very tired. There was no exhilaration in entering the hangar. He wondered what a hot shower would feel like. He wanted to get home. Then he and Torpey were walking down a hall and through the open door of the operations center. David Durham was at the desk. He hadn't gone home yet. Ted LeFeuvre looked at the clock. It was nearly half-past three.

Durham smiled when he saw them. His eyes were red-rimmed and he needed a shave.

"Welcome back, intrepid aviators," he said.

"Hello, Dave."

"You manage any sleep, sir?"

"Not much."

"You'll sleep good tonight."

"That would be different."

Durham explained how his crew had flown straight to the nearest shore, and then turned south and leapfrogged from one potential ditching site after another. They were never sure if they were going to make Sitka until they actually made the inside of the sound. When they landed at the air station, there was no more than twenty minutes of fuel in the 6029's tanks. The fuel warning light had been flashing.

Ted LeFeuvre left the situation report and the rest of the official paperwork to one of the watch hands. He signed what he had to and dated what he had to and then walked down to the locker room. He pulled off his flight suit and sat on a bench and looked at the lockers.

Bill Adickes walked in and said hello. He had not been home yet either. He said that he and Dan Molthen had gone out on a *second* sortie, with a different flight mechanic and rescue swimmer, to look for the fallen skipper and the lost, fifth fisherman. He said they had not stayed on scene long. Each time they tried to approach the water they had gotten a bad case of the shakes. Finally, they could not make their arms and hands move the flight controls so that they could go down close to the sea. So they pulled up and flew back to the air station.

"You see anybody in the water?"

"No, Captain," Adickes said. "We didn't."

"You're not going to believe who we pulled out of that mess," Ted LeFeuvre said.

"Who?"

"Bob Doyle."

"No."

"Yeah."

Adickes sat down on the bench. A clouded look washed over his face. Before his look had merely been one of exhaustion; now it was confused. His lips pursed.

"I just don't believe that."

"Believe it."

"You've got to be shitting me."

"No. I'm not. It was Doyle."

Adickes stood there and shook his head. He was trying to work through it in his mind.

"Did he—"

"He held those fishermen together."

Adickes nodded.

"If we hadn't sent him packing," Ted LeFeuvre said, "well . . . who knows?"

"Weird," Adickes said, "just too weird."

"Isn't it?" said Ted LeFeuvre.

When he walked in the house he did not turn on the light but put his duffel bag down on the floor and went to the den and looked at the pellet stove. It had gone out again. He went to the garage and dug out some pellets and returned to the den and restarted the stove.

In the dark, with the grandfather clock ticking and the paintings hanging on the walls exactly as he had left them and the glow of the streetlamps lighting the edges of the curtained windows, Ted LeFeuvre blew on his hands and waited. He thought about eating something. A sandwich, maybe. No. He was tired. Too tired even to do that. He stood there for a while and then went up the stairs to his bedroom.

He left the door open and turned on the lamp. The bed was still big and empty. Standing beside it, he undressed and put on his pajamas. Outside the heavy treaded tires of a Jeep or van rolling on the wet pavement went by and turned the corner. He lay down on the bed and pulled the covers over himself. The covers were cold. His nose was cold. That would change. The pellet stove was quite efficient. He reached over and turned off the lamp. Perhaps now he would be able to sleep.

He woke just before eight o'clock the next morning without need of an alarm. The house sat still and quiet. Outside it was raining. There was a heavy mist over the mountain. He remembered it was Sunday. He was supposed to do the reading at the morning church service. He looked at the clock. He had plenty of time to get ready.

He stood up and noticed the house was warm.

He went downstairs, checked the pellet stove, got more pellets from the garage, then went back upstairs and showered, brushed his teeth and hair and began shaving. He glanced in the mirror, put a small dab of cream on each cheekbone, then his chin and throat, and started scraping it off. He did not feel at all tired. Fourteen hours he had slept. Marvelous.

He raised his chin up, pulled it from side to side and went on scraping.

When was the last time he had slept that long? He could not remember. It must have been a while back. He ducked his face down to the sink, rinsed with cold water and toweled it. He looked at his face again carefully now in the glass. It sure looked like his face. That was his face all right. No different. After what he had just been through, it should be different, though, right? But no, nothing was different. Same old face. Same old Ted. He was back. Nothing had really changed, had it? He was still Theodore Cameron LeFeuvre of Whittier, California, age forty-six, with all of his faculties and flaws, all of his convictions and imperfections. Right? Right. And now it was Sunday and he was going to church and everything that had happened in the previous forty-eight hours, though undeniably real and certainly documented, seemed somehow like a story someone else had told.

He slipped on his beige sport coat, the one he always wore to church, a white, long-sleeved shirt, slacks, brown shoes, black, leather belt and striped tie and went downstairs. He locked the front door and stood on the porch.

The planter was empty.

Of course it is, he thought. It's winter. You don't put pansies outside in the winter. There is the greenhouse out back, though. Maybe he could plant something in the greenhouse. You haven't used the greenhouse yet. You could use it. Even in the winter.

With a lighter step, he walked over to the Cherokee, slid the key in the door lock. He climbed in, put the key in the ignition and looked in the rearview mirror.

Nothing was irrevocable. Nothing was forever. That was a significant thing. Only God was forever. The other business was not so tragic. None of it was important. Now that was a thought. That was a fine thought.

Still, it would be nice to have somebody to share this whole horrible thing with.

He turned the engine over and backed out of the driveway.

At the Trinity Baptist Church on Halibut Point Road, the congregation was seated and silent. They had just finished the hymn, and were waiting.

A slight, not-so-tall man with a gaunt, drawn face stood at the podium. He adjusted his wire-rimmed glasses and opened a book.

"If you would turn with me to Romans, chapter eleven," he said into the microphone, "I'll be reading verses thirty-three through twelve-two. I'll be reading from the New International Version."

The man paused.

"That's Romans, chapter eleven, verse thirty-three."

And he began to read:

> *Oh, the depth of the riches of the wisdom and knowledge of*
> *God!*
> *How unsearchable his judgments,*
> *and his paths beyond tracing out!*
> *Who has known the mind of the Lord?*
> *Or who has been his counselor?*
> *Who has ever given to God,*
> *that God should repay him?*
> *For from him and through him and to him are all things.*
> *To him be the glory forever! Amen.*

The reader took a breath, and continued:

> *Therefore, I urge you brothers, in view of God's mercy, to*
> *offer your bodies as living sacrifices, holy and pleasing to*
> *God—this is your spiritual act of worship.*
> *Do not conform any longer to the pattern of this world, but*
> *be transformed by the renewing of your mind.*
> *Then you will be able to test and approve what God's will*
> *is—his good, pleasing and perfect will.*

The man closed the book. There was not a murmur in the church. He sighed.

"May God add the blessing to the reading of His word."

And then Ted LeFeuvre walked over to his seat in the front row and got down on his knees and bowed his head. Those around him could only sit and wonder why, at that moment in the service, he was making such a display of reverence.

The next morning, Bob Doyle, Mike DeCapua and Gig Mork caught the first flight from Juneau to Sitka. DeCapua was antsy to get back. He did not sleep well in hotels, any place on land for that matter, and he had some accounts to settle up. Little John had advanced him a bag of weed for the fishing trip. The dope was going to cost him a hundred and twenty-five bucks, and he had not even smoked half of it. It had gone down with everything else on the ship.

"What are you going to do now?" Bob Doyle asked him.

"Try to get back on with the *Min E,*" DeCapua said. "Maybe Phil will cut me a break."

"Think he will?"

"If I grovel enough, I guess," DeCapua said. "Skippers like it when you grovel."

Bob Doyle nodded.

"Besides, you were the one who pissed him off with that stupid fucking note."

"I guess."

Gig Mork's mother met them at the airport. In the arrivals hall Bob Doyle looked around. He had wondered whether Tamara Westcott, the skipper's fiancée, might come out. He had thought of several things to say to her. He was going to tell her how sorry he was, other things. But she did not come. Maybe it's just as well, he thought. No need to make a scene in public.

The car was in the lot. They climbed in and headed out along the connector road, past the runway and the Eagle's Nest, and drove over the bridge. There was the gray of the channel, a small chop in the sound. It seemed like a very long ride. Nobody said anything as they rode.

"Let's get a beer at the P-Bar," Mork said. "Let's put our checks to work."

"Not today," Bob Doyle said.

"Come on," Mork said. "A Bud Light would do you some good."

"You go on ahead, Gig."

"Sure?"

"No," Bob Doyle said.

The car stopped out front of the Pioneer Bar and they got out. They closed the doors.

"Well, that's it," Mork said.

"See you around?"

"Sure."

They shook hands and Bob Doyle looked across the street and saw DeCapua, already on his way down the ramp of the ANB Harbor. He never had been one for send-offs. Bob Doyle stood on the street watching DeCapua until he was out of sight, and then he turned and started up the street.

The wind felt sharply cold and the clouds were pulling apart in the sky. He stopped and watched a brown paper bag roll and tumble and then get lifted into the air. Then he went on, feeling the wind lift his own hair, seeing it move the blades of grass in front of the Pioneer Home. He had lost his cap. He would get another. He saw someone he thought he knew, a mother, yelling at her child on the street corner. They're just having a bad day, he thought. They'll work things out. It did not matter. Nothing mattered. He kept on walking.

Along the main street everything looked new and changed. He had never seen the shops before. He had never seen the Sitka Hotel, or the bookshop, or the pelt store, or the pizza place, or the Columbia, or the cathedral. It was all strange and it was all different. It was like those days of late spring when he would walk home from school in Bellows Falls along a dusty track in the cool, moist shade of the pines, the distant sound of the cascading water of Little Egypt slobbering on the rocks, the scent of pine bark in his nostrils. He walked on, and it was as though his feet were not touching the ground. Everything was new. There was nothing to plan and nothing to think about. He could do all of that some other time. He did not have a thought in his head and he liked it that way.

It was like this walking past the Crescent Harbor. It was like this crossing Old Harbor Drive and it was like this passing Crescent Harbor and like this walking straight up the middle of Jeff Davis Street.

There was a light on in the kitchen. Georgia opened the door gave him a big bear hug.

"Oh, God," she said. "We heard what happened. Are you all right?"

"Fine," Bob Doyle said.

"Oh, you," she said. "Come in here, Bob. You probably want a shower, don't you? You must be hungry. A sandwich. How about a nice big ham sandwich? The kind you like."

"Sure."

"With lots of lettuce and tomato. Right. Tonight you stay in Robert's room. I've got clean sheets on the bed. Here, sit down. Sit down, yes, right here. You want a beer? Not yet? That's okay. Come, now. Just one. Yes. That's it. Let me get an opener for you. Now you tell us. Now you tell us everything."

Two weeks later, Bernice Honnold was in her living room on Lifesaver Drive running the vacuum cleaner when the doorbell rang. She turned the machine off and listened. The bell rang again. She went to the door.

"Hello," said a tall, lanky man with matted, red hair. He wore dirty jeans and a green cap. He had a nice, big smile. She had a feeling that she recognized him from somewhere. Behind him was another man, shorter, with a long black ponytail, a scar across his nose and blue eyes. He looked down whenever she looked his way.

"Yes?"

"Is Lee around?" the tall stranger said.

"No," she said. The man kept smiling. She smiled back. "Do you know him?"

"Oh, yes," he said. "I know him. I know him very well." The man looked down and saw the head of a girl, poking out from behind her mother's leg. "And who is this?"

"This is Hillary."

"Oh, hi, Hillary," said the man, and he crouched down to smile at her. "My name is Bob Doyle. And this here is Mike. Say, Hillary, did you know that your daddy is a hero, did you?"

"Yes," the girl said, and she giggled.

The man smiled. "I'm sure you did, Hillary. I'm sure of that." He cleared his throat, and then smiled again. "You know, you remind me a lot of somebody I know." He stood up. "Mrs. Honnold," he said, "when will Lee be home?"

"Oh, a little later." Bernice Honnold noticed the other man was holding a large sack in both arms. She wondered what was in the sack. "Maybe around four. He's standing duty today."

"Oh, that's a shame," said the man. "You see, we have something for him. And you, too."

"Oh?"

"We caught some rockfish and black bass and some crab, too. It's all fresh. All of it. We just caught it, you know. See, we're fishermen. And Mike, here, well, he even wanted to fillet them for you."

"Oh, my."

"See," said the man, "these are for you and your family. Would you mind if we left them in your kitchen?"

"I . . . uh," she stammered. "No, no, I suppose not. Of course not. Come in, please."

The quiet man with the long ponytail followed Bernice Honnold straight to the kitchen, pulled the fish out of the bag and laid them all out on the counter. He left the crabs in the sink. He began washing and cutting and separating out the bones.

"Those are some fish," she said.

"Yeah," the man said. "Mind if I smoke?"

"Of course not."

"Thanks." He rolled a cigarette and lit it.

The other man, the tall one with the red hair, said, "You know, I used to live here in Coast Guard housing, too."

"Really?" Bernice Honnold said. Now she recognized the man. He looked different now. Older. Much older. "In which house?"

"Across the street."

They chatted while the other man prepared the fish. He must be a cook, Bernice Honnold thought. But he doesn't say much. Not everyone does. Just look at the size of those fish, though. Where am I going to keep it all?

The taller man kept right on chatting.

Had she ever been to Kodiak? No, she said, but her husband was thinking of asking for a transfer there. Well, he said, I'm thinking of going to Kodiak myself. My kids are going there with my ex-wife. She asked him how old his kids were, their names. He told her.

"I don't want to be too far from them, you know."

"Of course not."

When the second man finished separating the two rockfish and big, black bass, he wrapped the fish in cellophane, stacked it neatly in the freezer, soaped and washed his hands and arms and toweled off quickly.

"This is really very nice of you," Bernice Honnold said.

The man just nodded.

"Well," the taller man said. "We're going to have to get going. We've got other people we need to visit. But you'll tell your husband we were here, won't you, Mrs. Honnold?"

"Oh, of course!"

"You'll tell Lee how grateful we are." He shook her hand.

"I will. I'll tell him."

"You don't know how grateful we are."

"I'll tell him."

"You just can't know."

"Are you sure you wouldn't want to stay awhile and tell him yourselves? He won't be but another hour or so."

"No," the man said. "We really ought to go. But you'll tell him for us, won't you?"

They were on the front stoop now. The second man, the quiet one, shook her hand.

He said, "Tell your husband thanks a bunch. He did a helluva job."

"I'll tell him."

"Great."

"Thank you, Mrs. Honnold," said the tall man. He shook her hand again, awkwardly. "Thank you again. Bye, Hillary! Bye, sweetheart! Bye!"

She watched them cross the lawn and turn on the sidewalk. She stood there in the doorway watching them. Then she felt a chill and closed the door softly.

The search for David Hanlon went full bore for ninety-four straight hours.

Seven H-60 helicopters from Sitka and Kodiak, three Kodiak C-130 jets and two cutters, the *Planetree* and the *Anacapa*, combed an area of ocean twice the size of the state of Rhode Island. The planes took off at first light and returned after dusk. The cutters put their lights on and searched nights, too.

At 10:05 A.M., on Sunday, February 1, 1998, the *Planetree* spotted a gray fish tote bobbing in the waves at 58°43.14' north, 139°05.01' west. Ten minutes later, it pulled up a four-foot-square, white piece of wood nearby, then an orange tarp. Finally, at 2:58 P.M., it plucked out a lone, white life ring with the words LA CONTE painted on it in black lettering.

But there was no sign of Hanlon.

After heavy fog and darkness set in, the *Planetree* picked up a 406-megahertz hit at 58°44.4' north, 138°58' west. It raced to the coordinates and retrieved an orange EPIRB, its ring switch set to ON and still transmitting weakly.

A check of the serial number confirmed that the beacon had come from the *La Conte*. Six hours later, the cutter recovered a second EPIRB, a 121.5-megahertz model.

During the next two days, with aircraft flying grid after grid northwest of the Triple Forties, helicopter crews sighted a white deck cover, part of a bare wood deck, two white marker floats, another fish tote and a two-by-two piece of plywood.

But no survivor, no body.

On Monday morning, February 2, a C-130 pilot spotted a three-foot-long orange object floating in the waves. But he lost sight of it and a series of subsequent helicopter sorties turned up nothing. The orange object was not seen again.

For the next day and a half, the *Planetree* crisscrossed the waters north and west of the Fairweather Grounds, using searchlights and infrared scanners. On Monday night the weather deteriorated; winds rose to thirty knots and seas to twenty feet, and visibility diminished considerably.

The search was finally called off at five in the afternoon on Tuesday, February 3, $678,545 after it had begun.

People reckoned it would not be more than a week before a crab boat spotted the missing fisherman or a seiner scooped up his remains in a net. But then a week went by, and another week, and then another. Then a month passed. Then six months.

For David Hanlon's five sisters and two brothers, life began to drift.

Without a body, it was difficult to let him go.

There was a formal Coast Guard inquiry, of course.

Three days of public hearings were held in Juneau and in Sitka a week after the incident, and the Coast Guard appointed a special investigator, Lieutenant Commander David C. Stalfort, to head the probe. He took 523 pages of testimony from fourteen people, including the ship's two previous owners, shipwrights, former crewmen familiar with the *La Conte* and the three survivors and thirteen airmen who took part in the rescue.

In the end, Stalfort concluded that a "catastrophic event occurred that allowed uncontrolled flooding into the hull," but that the "cause or nature of this event is unknown." His final report noted that the owner "did not maintain the vessel in a condition that provided adequate strength for conditions likely to be encountered offshore in the Gulf of Alaska in the winter."

It said casualties "could have been prevented had the master, Mark Morley, heeded the weather warnings," made sure to provide a life raft, a single sideband radio and personal marker lights for their survival suits and trained the crew to use all of the bilge pumps on board in case of an emergency.

"It is recommended that there be no further investigation of the owner, Scott Echols, for criminal negligence."

He recommended that the investigation be closed and it was, without delay.

EPILOGUE

It was a quarter past six and the investigators' office of the Alaska state crime lab was quiet except for the sound of a man in a corner cubicle pecking at the keyboard of a desktop computer. Everyone else had gone home. It was August 31, 1998, which happened to be David Hanson's seventh wedding anniversary, and he was anxious to get home, too, to take his wife, Valery, out to dinner. But he had a report to finish. Nobody had actually told him to put a rush on it. It was just a missing persons file. But to him it was important, somehow, that he do it right away.

So he kept typing.

Three hours earlier, Walter MacFarlane had brought startling news: the fingerprint lab had matched a print from a fisherman named David Hanlon to another they had reproduced from a skin fragment found inside a bear-mauled survival suit on Shuyak Island. The word *match* had hit him like a stiff, electric jolt; after nineteen days trying to identify the remains, he, David Hanson, the rookie investigator, was going to stroll into his sergeant's office and declare that he'd cracked his very first case.

He hadn't strolled—he'd bounded, as a matter of fact, over to his boss's door and rapped on it. Sergeant Mike Marrs, hunched over a stack of papers, looked up. It was a grumpy look.

"It's him," Hanson said.

"Who?"

"Hanlon."

He held out the fingerprint card. Marrs, still eyeing him moodily, took it. He moved his dark eyes up and down the card, over the fingerprints rolled in black ink. Hanlon's right index finger was circled in red. That was the finger that had been positively matched to the skin found in the neoprene mitten on Shuyak.

Marrs removed his thick glasses and flicked them on his desk. He looked Hanson straight in the eyes. The sergeant's annoyed expression had melted away.

"Good work, young man," Marrs said almost gently. Then he gave Hanson a small, approving smile.

"Thanks."

"Now," Marrs had said, "it's time to go show the captain. You want to do the honors?"

Once they had notified all of the higher-ups, Hanson had telephoned the troopers' station in Angoon and asked that an officer go in person to Hoonah to tell the Hanlons. That wasn't something you did on the phone, he had said to himself. I'm sure there must be a relative that can break the news gently to that family. Well, at least they could grieve and start putting David's death behind them. All of the Hanlons could rest.

So now all that was left was to finish typing up his final report on the case. At one point, he noticed a mistake in his narrative:

the deceased, David Hanson, was

He hit the backspace key, deleted the word *Hanson*, retyped the name so that it correctly read *Hanlon* and went on.

He had completed another page when again he noticed, in another reference to the dead man, that he had typed:

David Hanson

He stared at his mistake.

Why do I keep doing that? I need to hit the L key, not the S key. Okay, erase that and type H-A-N-L-O-N. There you go. That's better.

While Hanson was proofreading his report he noticed yet another sentence where his own name had been entered instead of the name Hanlon. And then it hit him.

David Hanlon. *The mystery man I've been trying so hard to identify has a name that's just* one *letter different from my own. We're not all that far apart, are we, Mr. Hanlon?*

He shuddered and hit the delete key.

On April 2, 1998, four of the aviators from the third helicopter rescue team—Ted LeFeuvre, Steve Torpey, Fred Kalt and Lee Honnold— received the Distinguished Flying Cross, the highest aviation honor given in the United States during peacetime. Mike Fish, the team's rescue swimmer, was awarded the Coast Guard's Air Medal.

Later, all five airmen received the National Association of Naval Aviation's Outstanding Achievement in Helicopter Aviation, the Naval Helicopter Association's national award for helicopter rescue, the Naval Helicopter Association's Regional prize, the Aviation Week and Space Technology Hall of Fame Laureate and *Rotor and Wing* magazine's award for heroism in helicopter aviation.

The crews of the other two rescue helicopters received Coast Guard commendation medals, achievement medals and letters of commendation for their efforts.

At the conclusion of his Sitka tour in June of 1998, Ted LeFeuvre became chief of the Search and Rescue Branch of the Seventh Coast Guard district in Miami, Florida. During the next two years he oversaw seventeen thousand search-and-rescue cases and helped develop a new Coast Guard aviation program that led to the seizure of seven tons of illegal drugs worth an estimated $120 million. For his efforts he received the Meritorious Service Medal in 2000 for exceptional leadership.

From 2000 to 2003, Ted LeFeuvre served as commander of Air Station Humboldt Bay in Eureka, California, where he commanded rescue operations of that state's northernmost coastal region. In June 2003, he was appointed the Coast Guard's liaison officer to the Naval Education and Training Command in Pensacola, Florida.

In 1999, while attending a Sunday church service, he met a school-teacher in Miami, Renee Browning. After dating for five months, the couple wed on April 1, 2000. They remain happily married. Michelle and Cam LeFeuvre now live in Atlanta. Cam is a mechanic in a Volkswagen dealership, and Michelle, who works at a law firm, has patched things up with her father. She and her younger brother frequently visit their father in Pensacola.

David Durham was transferred in May 2000 to San Diego, where he served as a deputy commander. In June of 2002, he was given his first command: Air Station Sitka. Today he works and lives in Sitka with his wife, Trish.

Steve Torpey was promoted to commander in the Coast Guard and is studying for a master's degree in organizational management. After serving as a flight instructor at the Aviation Training Center in Mobile, Alabama, he was reassigned to Air Station Clearwater, where he flies H-60s and serves as an assistant operations officer. He and his wife, Kari, live in Trinity, Florida, with their five-year-old son and three-year-old daughter.

Russ Zullick was transferred to Kodiak, Alaska, in September 1999 and promoted to lieutenant commander. In Kodiak he earned his third, fourth and fifth Coast Guard Achievement Medals and a Distinguished Flying Cross for rescuing two survivors whose small aircraft crashed into the side of almost inaccessible mountain along the Alaska Peninsula. He lives in Kodiak with his wife, Deborah, and their three-year-old son.

Dan Molthen lives with his wife, Theresa, and their three children in Camden, North Carolina, and is a lieutenant commander at Air Station Elizabeth City. On December 17, 2000, the cruise ship *Sea Breeze* I began taking on water more than two hundred miles off the Carolina coast with thirty-four people on board. Molthen and his crew hoisted twenty-six survivors off the ship, crammed them into the Jayhawk's six-and-a-half-by-eight-foot cabin and somehow made it safely back to base. The rescue set a new record for number of people flown in an H-60. A second rescue helicopter picked up the remaining eight survivors.

Bill Adickes left Alaska in 1998 with his wife, Carin, and their two sons and went to work as a C-130 pilot in Sacramento, California. After a four-

year tour, Adickes was reassigned to Hawaii and now flies fishery patrols in
the Pacific, often between Guam and Midway. He has put in for a transfer
back to Alaska; he says he misses the people of Sitka, the wilderness and
the excitement of helicopter rescues.

Fred Kalt, Lee Honnold, Chris Windnagle and Sean Witherspoon still
work as flight mechanics.

Kalt is a chief in Elizabeth City; Windnagle is a chief in Kodiak. After
a stint in Cape Cod, Massachusetts, Witherspoon was reassigned to
Clearwater, Florida. Honnold is now a flight engineer on C-130 planes
and based in Kodiak.

Rich Sansone went on to Officer Candidate School and recently was
commissioned as an ensign in Cleveland, Ohio. A. J. Thompson continues
as a rescue swimmer, now based at Air Station Elizabeth City in North
Carolina. Mike Fish quit the Coast Guard shortly after the *La Conte* case
and became a fireman. He lives in Anchorage now with his wife, Heidi,
and two children.

David Hanson remains an investigator at the state crime lab in
Anchorage, and lives in nearby Eagle River with his wife, Valery, and their
six children. On February 1, 2004, he was appointed chief investigator of
the state's Missing Persons Bureau. The big map with all of the colored
pins indicating where people have disappeared in Alaska is now his responsibility.

David Hanlon's remains were shipped to his family in Hoonah, and on
September 10, 1998, some two hundred people attended a memorial service for him at the Bethel Christian Center. Honoring an ancient Tlingit
custom, his relatives later sailed out into the Icy Strait and sprinkled white
wreaths, daisies and carnations across the water's surface. He is buried at
the Evergreen Cemetery in Juneau, where his mother and father were laid
to rest.

Mark Morley's body was flown back to Michigan, where a service was
held at the Uhts Funeral Home in Westland. He was buried in nearby Ply-
mouth on February 6, 1998—four days short of his thirty-sixth birthday.

That April, his fiancée, Tamara Westcott, had her last name legally
changed to Morley. On August 13, 1998—at exactly the same hour that
Jesse Evans and Doug Conner came upon David Hanlon's remains on
Shuyak Island—she gave birth to an eight-pound, eleven-ounce boy. She

named him Mark Morley Jr. She still lives in Sitka with her daughter, Kyla, and her son.

When he is not fishing or repairing boats, Gig Mork lives with his parents in a new Native housing development out near Indian River in Sitka. His reputation as a solid deckhand on boats—and a solid drinker in taverns—has not suffered as a result of the *La Conte* episode.

In the aftermath of the sinking, Mike DeCapua returned to longlining. After an account of the *La Conte* sinking appeared in *Reader's Digest* in 1999, his youngest daughter, Melanie Thistle, contacted him. (As it turned out, she had been trying to locate her father for years.) Not long after, Mike DeCapua flew to Spokane and spent two weeks with his daughters—his first visit in two decades.

On his forty-sixth birthday, July 19, 2002, Mike DeCapua telephoned his older brother, Don, with news: he had met a wonderful woman named Wendy and had driven with her from Sitka to Michigan in her car. Two days later, Don DeCapua got another call—this time from Wendy. She and Mike had apparently gone out to have dinner and drinks, separated and planned to meet back at her car, but Mike never showed up. Since he had left all of his personal belongings at her house, Wendy called the police two days later to file a missing persons report. Six months passed with no word. Then, last January, Mike called his brother from Tampa, Florida, where he apparently had been fishing.

Soon after, Mike DeCapua returned to Sitka and his life of fishing on the high seas.

Bob Doyle's ex-wife, Laurie, married Rick Koval on May 1, 1998, and the two moved with Brendan and Katie to Kodiak. In August 1999, she and Koval divorced. Laurie and the kids still live in Kodiak.

Bob Doyle traveled to Kodiak twice to visit his children, and then moved back to Bellows Falls. He began working odd jobs, including painting homes, working in a meatpacking house, catering, and bartending at Nick's, the bar and pool hall where he tipped his first beer as a fifteen-year-old.

He is now dating a local nurse and lives with his younger sister, Sally, at their grandparents' old home in North Walpole, across the Connecticut River from Bellows Falls, in New Hampshire. To this day he keeps a snapshot of Mark Morley in his wallet.